Functional Bladder Reconstruction Following Spinal Cord Injury via Neural Approaches

Chunlin Hou
Editor

Functional Bladder Reconstruction Following Spinal Cord Injury via Neural Approaches

Springer

Editor
Chunlin Hou, M.D
Department of Orthopedic Surgery
The Second Military Medical University
Changzheng Hospital
Shanghai
People's Republic of China

ISBN 978-94-007-7765-1 ISBN 978-94-007-7766-8 (eBook)
DOI 10.1007/978-94-007-7766-8
Springer Dordrecht Heidelberg New York London

Printed on acid-free paper

Springer is part of Springer Science+Business Media (www.springer.com)

Preface

Spinal cord injury (SCI), either traumatic or non-traumatic, is a devastating disorder that leads to significant physical, psychological and socioeconomic burden on patient and family. Spinal cord injury often occurs in young people, who have the prospect of an almost normal life expectancy, but a considerably impaired quality of life. These patients not only experience severe dysfunctions of voluntary movement of limbs, but also have impaired functions of pelvic organs, such as the bladder and the bowel, and sexual dysfunction. In practice, one of the most distressing aspects of spinal cord injury is not being able to regulate bladder function: urine reservoir and voiding.

Managing urinary functions in spinal cord–injured patients remains a challenge for clinical medical workers as well as basic researchers. Generally, the options rely on indwelling or intermittent catheterization, drugs, operations and other medical devices. Our group has performed experimental study and clinical investigations to improve bladder function after spinal cord injury for about 20 years, mainly focused on neural approaches. We have received six research supports from national Natural Science Foundation of China (NSFC). We have published more than 100 papers, in Chinese and/or English. Our series study has also won first prize from the Chinese Medical Association.

This book is a culmination of the author's investigation that attempt to learn more about bladder function in spinal cord injury. After a brief review of spinal cord injury and related anatomy, the main topic of this book is focused on functional bladder reconstruction through neural approaches, including somato-CNS-bladder artificial reflex arc establishment through neural anastomosis, bladder denervation through selective sacral root rhizotomy and sacral root stimulated micturition through electrical device implantation. Our intent is to stimulate passion for further research in this field and to enable improved care and outcomes for individuals with spinal cord injury.

Shanghai, People's Republic of China Chunlin Hou, M.D.

Contents

An Overview of Traumatic Spinal Cord Injury

1

Shimin Chang and Chunlin Hou

Spinal cord injury (SCI), either traumatic or non-traumatic in origin, can lead to significant physical, psychological and socioeconomic burden on patient and family. Since SCI causes extensive lifelong consequences, epidemiological data are of fundamental importance in tracing its occurrence, deciding upon preventive strategies, and planning clinical resources and social services. However, there is a lack of accurate data on the epidemiology of spinal cord injuries in many countries. Although some patients of SCI are caused by non-traumatic reasons, such as, tumor, tuberculosis, spine degeneration, and so on, traumatic spinal cord injury (TSCI) remains the mainstay of epidemiologic survey of SCI.

1.1 Prevalence of Traumatic SCI

The range of reported global prevalence is between 280 and 4,187 per million [1]. The prevalence of TSCI by country and authors of published data are presented in Table 1.1.

S. Chang
Department of Orthopedic Surgery, Yangpu Hospital, Tongji University School of Medicine, Shanghai, People's Republic of China
e-mail: shiminchang11@aliyun.com

C. Hou, M.D. (✉)
Department of Orthopedic Surgery, Changzheng Hospital, The Second Military Medical University, Fengyang Road 415, Shanghai 200003, People's Republic of China
e-mail: chunlin_hou@163.com

1.2 Incidence of Traumatic SCI

A global-incident rate is estimated at 23 TSCI cases per million (179,312 cases per annum) in 2007 [1]. Regional data are available from North America (40 per million), Western Europe (16 per million) and Australia (15 per million). Extrapolated regional data are available for Asia-Central (25 per million), Asia-South (21 per million), Caribbean (19 per million), Latin America, Andean (19 per million), Latin America, Central (24 per million), Latin America-Southern (25 per million), Sub-Saharan Africa-Central (29 per million), Sub-Saharan Africa-East (21 per million).

The most representative incidence statistic for each country within WHO global regions is presented in Table 1.2, along with available etiology data, level of injure and survival rate.

Directly interpretable incidence data has not changed obviously in the high income regions such as North America, Western Europe and Australia. Because of the economic limitations, it is difficult to acquire accurate epidemiological data for TSCI in many developing countries. But it is common accepted that in the emergent countries such as in China, the incidence and prevalence of TSCI are increased steadily. The reasons include: (1) the injury cases due to high fall and heavy crush increased parallel to economy development; (2) as the population ages, the number of TSCI in elderly patients also increases; (3) as the economy develops, more patients with TSCI be recorded and registered normally, and they were

C. Hou (ed.), *Functional Bladder Reconstruction Following Spinal Cord Injury via Neural Approaches*, DOI 10.1007/978-94-007-7766-8_1, © Springer Science+Business Media Dordrecht 2014

Table 1.1 Prevalence of TSCI by country and author(s) of published data (From Lee BB, et al. Spinal Cord. 2013 [1], with permission)

Country	Author(s) of published data	Observation period	Prevalence per million population
Australia	O'Connor [2]	1997	681
Finland (Helsinki)	Dahlberg [3]	1999	280
Norway (Hordaland and Sogn og Flordane)	Hagen [4]	2002	365
Iran (Tehran)	Rahimi-Movaghar [5]	2008	440
Canada	Rick Hansen SCI Register [6]	2010	1,184
USA	Cahill [7]	May 2008–August 2008	4,187

missed in the past. With 60,000 new cases a year, China has more patients with spinal-cord injury than anywhere else in the world [15].

1.3 Injury Site, Age and Cause of TSCI

The most common site of TSCI was the cervical area, followed by the thoracic. Cervical, thoracic, lumbar and sacral lesions on average accounted for 54 %, 24 %, 19 % and 1 % of the cases, respectively [16].

For age-specific TSCI incidences, most studies described a bimodal distribution [16]. The first peak was demonstrated in young adults, in an age group between 15 and 29 years. The second peak was observed in elderly people aged 65 years and older. Elderly people are more prone to injury because of degenerative spine conditions, such as stenosis, spondylolisthesis, and degenerative disc disease. As the population ages, the number of TSCI in elderly patients also increases.

In TSCI, the sex ratio ranged considerably, from 2:1 in Turkey to 6.7:1 in Ireland. In contrast, a lower sex ratio of 1.3:1 was reported for non-traumatic SCI.

With regard to cause [17], traffic accidents are the most common cause of TSCI in North America, Europe and sub-Saharan Africa. In western Norway, falls were the most common cause. In Asia, motor vehicle collisions and falls were the main causes. It is not surprising that the proportion of new TSCI due to falls has been increasing steadily as Asian societies age. The relative distribution of causes is influenced by lifestyle and economic environment. For example, in some regions of Turkey, the most common cause of TSCI was falling from a great height, which could be associated with the fact that most falls occurred in the summer when people sleep on the top of their houses. In China, there was an enormous number of elderly patients with degenerative cervical spine changes who were more vulnerable to a slight damage (e.g. a low fall), so the number of low falls-induced TSCI would increase gradually; in Bangladesh, falls while carrying heavy loads on neck or back is a common cause. In Afghanistan, Jordan, and southeast Turkey, the frequency of firearm wounds, a specific problem to these countries, was higher than in other regions. In Jordan, the majority of firearm accidents happened during wedding celebrations, a common cultural habit in rural areas. In other regions, where the political situation was stable and firearms were strictly controlled, such injuries were rare.

1.4 Life Expectancy After Traumatic SCI

A cohort study in Sydney [18] from 1995 to 2006 showed that, from 2,014 persons, 88 persons with tetraplegia (8.2 %) and 38 persons with paraplegia (4.1 %) died within 12 months of injury, most often with complete C1-4 tetraplegia. Among first-year survivors, overall 40-year survival rates were 47 % and 62 % for persons with tetraplegia and paraplegia, respectively. The most significant increases in mortality were seen in those with tetraplegia and American Spinal Injury Association Impairment Scale (AIS) grades A-C lesions, with Standardized mortality ratios between 5.4 and 9.0 for people <50 years, reducing with advancing attained age. Estimated life expectancies

Table 1.2 Incidence, neurological level and aetiology of TSCI by Region and Author(s) of published data

Country or region	Author(s)	Observation period	Incidence per million population per year	Neurological level of injury		Aetiology of SCI	Survival		Comments
				Tetraplegia %	Paraplegia %		1-year mortality	10-year mortality	
China (Tianjin)	Ning [8]	2004–2008	23.7 (n=869; males, n=738, females, n=131)	72 %	28 %	Land transport (34 %, n=296), falls (57 %, n=495), sports/ recreation (0.2 %, n=2), violence (1.4 %, n=12), work-related (0.8 %, n=7) and other/ unknown (6.5 %, n=57)	1.4 % (n=12)		All deaths occurred during hospitalization, patients had cervical TSCI. Among the 12 patients, 42 % (n=5) were 60 years or older, 10 were males and 11 died of respiratory failure
China (Beijing)	Li [9]	2002	60.6 est. (n=1,079; n=264; males, n=198, females, n=66)	5 %	95 %	Land transport (22 %, n=59), falls (41 %, n=109), sports/ recreation (1 %, n=3), violence (0.4 %, n=1) and other/unknown (35.6 %, n=94)			
China Taiwan	Chen [10]	1992–1996	18.8 (n=1,232; males, n=958, females, n=274) (Age group 15–64 years)	43 %	57 %	Land transport (49 %, n=609), falls (41 %, n=500), sports/ recreation (1 %, n=16), violence (2 %, n=25) and other/ unknown (7 % n=82)	6 % (n=73)		Of those who died in hospital, 40 died of respiratory failure, 22 of associated injuries, 7 died of complications (renal failure, pressure sore, gastrointestinal bleeding) and 3 committed suicide

(continued)

Table 1.2 (continued)

Country or region	Author(s)	Observation period	Incidence per million population per year	Neurological level of injury		Aetiology of SCI	Survival		Comments
				Tetraplegia %	Paraplegia %		1-year mortality	10-year mortality	
China Taiwan	Chen [10]	1992–1996	47.5 (Geriatric population, n=297; males, n=186, females, n=111) (Age group 65-and above years)	54 %	45 %	Land transport (33 %, n=97), falls (62 %, n=185), sports/recreation (0.3 %%, n=1), violence (2 % n=5) and other/unknown (2.7 % n=9)	11 % (n=33)		Of the 33 elderly patients who died in hospital, 17 died of respiratory failure, 12 of complications and 3 committed suicide
Australia	Norton [11]	2007–2008	15.0 (Age-standardised); (n=285; males, n=249, females, n=45)	53 %	47 %	Land transport (46 %, n=130), falls (28 %, n=81), sports/recreation (13 %, n=37) and other/unknown (13 %, n=37)			
Norway (Hordaland and Sogn og Flordane)	Hagen [4]	1952–2001	26.3 (Annual incidence in 1997–2001; n=71); (n=336; males, n=277, females, n=59)	52 %	48 %	Land transport (34 %, n=115), falls (46 %, n=153), sports/recreation (9 %, n=29), violence (1 %, n=9) and other/unknown (10 %, n=34)	21 % (n=83)	10.8 % median (Range 7.4 %–33.3 %)	First year mortality differences may be due to a combination of societal variables (car use, age of population) and emergency medicine procedures. Suicide was relatively high. Main causes of death after TSCI were: cardiovascular (39 %), respiratory disease (35 %), neoplasm 17 %, nervous system disease (17 %), genitourinary disease (12 %), suicide and accidental poisoning (6 %)

Sweden (Stockholm)	Divanoglou [12]	2006–2007	19.6 (n=49; males, n=36, females, n=11 [consenting cases])	45 %	55 %	Land transport (23 %, n=11), falls (38 %, n=18), sports/ recreation (17 % n=8), self-harm (9 %, n=4), assault (2 %, n=1) and other/ unknown (11 %, n=5)	0 % deaths in Stockholm compared to 18.8 % (n=12) in Greece			Major cause of deaths in Thessaloniki was pneumonia +/– sepsis (50 %) followed by UTI and cardiac arrhythmias
Canada (Ontario)	Furlan [13]	2007	38 (n=487; males, n=408, females, n=79)	70 %	30 %	Land transport (49 %, n=240), falls (20 %, n=95) and other/ unknown (31 %, n=152)	8 % (n=37)			The vast majority of deaths in the acute and chronic stages following SCI were attributed to cardiac arrhythmias (37.8 %), respiratory complications (37.8 %) and coagulation-related complications (10.8 %)
USA	NSCISC [14]	2005	40 (n=11,000 approx.; males, n=8,756, females, n=2,244)	53 %	47 %	Land transport (48 %, n=5,280), falls (23 %, n=2,530), sports/ recreation (9 %, n=990), violence (14 %, n=1,540) and other/unknown (6 %, n=660)				

from 25 to 65 years ranged between 69–64 %, 74–65 %, 88–91 % and 97–96 % for C1-4 AIS A-C, C5-8 A-C, T1-S5 A-C and all AIS D lesions, respectively.

1.5 Prevention of SCI

1.5.1 Primary Prevention

This is mandatory in many developed countries that participants should receive professional education before they take part in some activities such as bicycle riding, diving, high-altitude operations. Other countries may lack training, tools or financial resources to undertake this task. The governments and non-governmental organizations of developing countries should increase financial resources to prevent TSCI. The experience of developed countries demonstrate that it is important to popularize the knowledge of people cope with nature disaster, such as earthquake and mud-rock flow, and this can decrease the cases of TSCI.

1.5.2 Secondary Prevention

Refer patients with TSCI to a specialized centre, equip the specialized centers with essential resources to accept referrals, training physicians with internationally accepted data standards and management program, are important secondary prevention methods of TSCI.

1.5.3 Tertiary Prevention

The tertiary prevention of TSCI include, complications prevention, rehabilitation program, and necessary functional reconstruction surgery. The most urgent demand of most paraplegia patients is urination. Bladder function reconstruction is vitally important to enhance the quality of life in patients with SCI.

References

1. Lee BB, Cripps RA, Fitzharris M. The global map for traumatic spinal cord injury epidemiology: update 2011, global incidence rate. Spinal Cord. 2013. doi:10.1038.
2. O'Connor PJ. Prevalence of spinal cord injury in Australia. Spinal Cord. 2005;43:42–6.
3. Dahlberg A, Kotila M, Leppanen P, Kautiainen H, Alaranta H. Prevalence of spinal cord injury in Helsinki. Spinal Cord. 2005;43:47–50.
4. Hagen EM, Eide GE, Rekand T, Gilhus NE, Gronning M. A 50-year follow-up of the incidence of traumatic spinal cord injuries in Western Norway. Spinal Cord. 2010;48:313–8.
5. Rahimi-Movaghar V, Saadat S, Rasouli M, Ganji S, Ghahramani M, Zarei M-R, et al. Prevalence of spinal cord injury in Tehran, Iran. J Spinal Cord Med. 2009;32:428–31.
6. Farry A, Baxter D. The incidence and prevalence of spinal cord injury in Canada: overview and estimates based on current evidence. Rick Hansen Institute and Urban Futures Institute, December 2010.
7. Cahill A, Fredine H, Zilberman L. Initial briefing: prevalence of paralysis including spinal cord injuries in the United States, 2008. Christopher and Dana Reeve Foundation, Paralysis Resource Foundation, The University of New Mexico, School of Medicine; 2009.
8. Ning GZ, Yu TQ, Feng SQ, Zhou XH, Ban DX, Liu Y, et al. Epidemiology of traumatic spinal cord injury in Tianjin, China. Spinal Cord. 2011;49:386–90.
9. Li J, Liu G, Zheng Y, Hao C, Zhang Y, Wei B, et al. The epidemiological survey of acute traumatic spinal cord injury (ATSCI) of 2002 in Beijing municipality. Spinal Cord. 2011;49(7):777–782.
10. Chen H, Chen S, Chiu W, Lee L, Hung C, Hung C, et al. A nationwide epidemiological study of spinal cord injury in geriatric patients in Taiwan. Neuroepidemiology. 1997;16:241–7.
11. Norton L. Spinal cord injury, Australia 2007–08: injury research and statistics series no. 52. Cat. no. INJCAT 128. Canberra: AIHW; 2010.
12. Divanoglou A, Levi R. Incidence of traumatic spinal cord injury in Thessaloniki, Greece and Stockholm, Sweden: a prospective population-based study. Spinal Cord. 2009;47:1–6.
13. Furlan JC, Bracken MB, Fehlings MG. Is age a key determinant of mortality and neurological outcome after acute traumatic spinal cord injury? Neurobiol Aging. 2010;31:434–46.
14. National Spinal Cord Injury Statistical Center. Spinal cord injury. Facts and figures at a glance. J Spinal Cord Med. 2005;28:379–80.

15. Qiu J. China spinal cord injury network: changes from within. Lancet Neurol. 2009;8(7):606–7.
16. van den Berg ME, Castellote JM, Mahillo-Fernandez I, de Pedro-Cuesta J. Incidence of spinal cord injury worldwide: a systematic review. Neuroepidemiology. 2010;34:184–92.
17. Ning GZ, Wu Q, Li YL, Feng SQ. Epidemiology of traumatic spinal cord injury in Asia: a systematic review. J Spinal Cord Med. 2012;35(4):229–39.
18. Middleton JW, Dayton A, Walsh J, Rutkowski SB, Leong G, Duong S. Life expectancy after spinal cord injury: a 50-year study. Spinal Cord. 2012;50:803–11.

2 Clinical Anatomy of Cauda Equina and Lumbosacral Nerve Root

Ruisheng Xu and Chunlin Hou

2.1 Introduction

Treatment of patients with neurogenic bladder dysfunction after spinal cord injure (SCI) is a challenge to the doctors all over the world till now. Although there are a lot of experimental new ways for recovery of neurogenic bladder dysfunction, it's still far from clinical application [1]. Shanghai Changzheng hospital developed a new method for SCI patients' bladder functional reestablishment with Achilles tendon reflex, which makes a certain use of somatic reflex under SCI level to control voiding [2]. The "tendon-CNS(central nerve system)-bladder" route is connected by ventral spinal roots cross anastomosis, then somatic motor impulses is passed to bladder smooth muscle through this reflex, so as to initiate voiding by stimulate Achilles tendon reflex [2]. As experimental study achieved satisfactory outcomes from dog SCI models [3, 4], anatomical study should be done before the step of more clinical applications. So the authors observed the component of cauda equina and lumbosacral nerve root to provide anatomical bases for the choice of proper spinal roots, the idea level for cross anastomosis and the identification of ventral spinal roots during the operation of bladder functional reconstruction with Achilles tendon reflex, knee reflex or lower abdominal reflex.

R. Xu
Department of Orthopedic Surgery, Wuxi No.3 People's Hospital, Jiangshu, People's Republic of China
e-mail: xuruisheng@medmail.com.cn

C. Hou, M.D. (✉)
Department of Orthopedic Surgery, The Second Military Medical University, Changzheng Hospital, Fengyang Road 415, Shanghai 200003, People's Republic of China
e-mail: chunlin_hou@163.com

2.2 Materials and Methods

All the date were observed and measured from 20 adult formalin-fixed specimens (male: 14, female: 6, 40 sides; donated voluntarily). Observation was focused on the anatomy of nerves which related to knee jerk reflex, Achilles tendon reflex and reflex of lower abdominal wall.

Firstly, posterior aspect of T_{11}–S_1 spinal roots and sacral plexus was exposed clearly, then the spinal roots which origin lumbar plexus, sacral plexus was observed (Fig. 2.1). The diameters of T_{11-12}, L_{2-5} and S_1 spinal roots were measured in the position where it passes through intervertebral foramen rightly and the diameters of the branches that composes sciatic or sacral plexus nerve was measured and recorded, then back wall of canalis veterbrale and sacralis, spinal dura were opened, and medullary conus, lumbar and sacral spinal roots were showed. Then a photo was taken when the end point of medullary conus was marked (Fig. 2.2), relative position between each ventral and dorsal roots was recorded, distance between $T_{11}T_{12}$ to $S_2 \sim S_4$ ventral roots, as well as the overlapping length between $L_2 \sim S_1$ to $S_2 \sim S_4$ ventral roots, diameters and horizontal area of each root

C. Hou (ed.), *Functional Bladder Reconstruction Following Spinal Cord Injury via Neural Approaches*,
DOI 10.1007/978-94-007-7766-8_2, © Springer Science+Business Media Dordrecht 2014

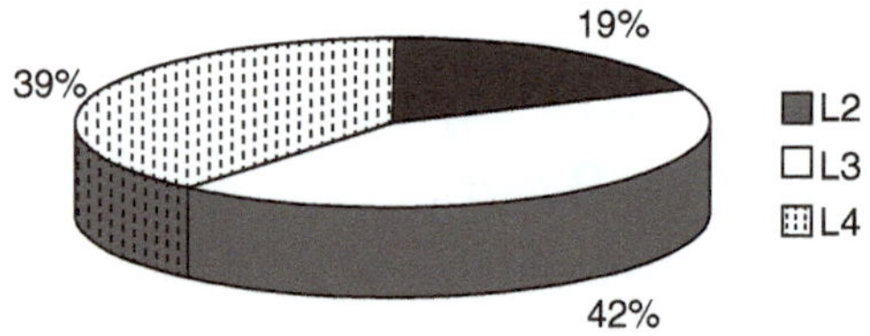

Fig. 2.1 L2, L3 and L4 spinal roots' contribution to femoral nerves

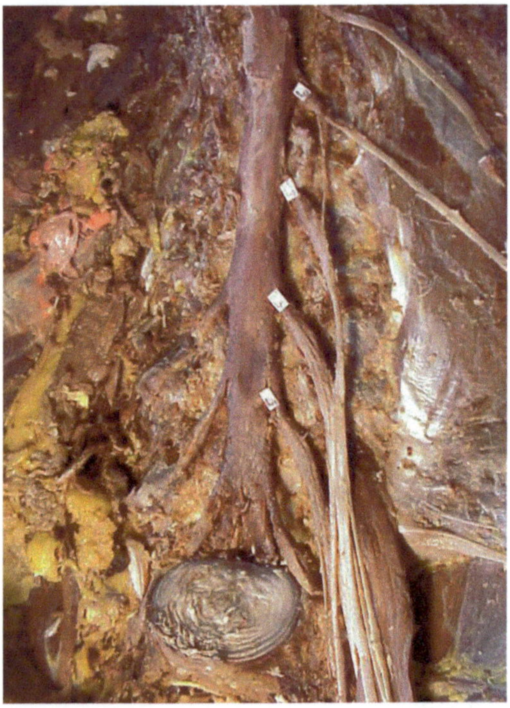

Fig. 2.2 Component of femoral nerves—L2, L3 and L4 spinal roots

was measured and calculated with a sliding gauge (degree of accuracy: 0.02 mm) and a stainless steel rule (accurate degree: 1 mm).

2.2.1 Anatomical Study of Knee Jerk Reflex Related Nerve: Femoral Nerve

Firstly, femoral nerve was tracked to its' central terminal, and the contribution of L_2, L_3 and L_4 spinal nerve roots to the femoral nerve was observed. The diameter of each root at the point of intervertebral foramen was measured, and the diameter of each branch contributing to femoral nerve was recorded as well. Then, a picture was taken after the exposure of conus medullaris and cauda equina, as the end of spinal cord cone was marked with a needle, the anterior-posterior and sequence relationships of L_{2-4} and S_{2-4} anterior and posterior roots were recorded.

2.2.2 Anatomical Study of Achilles Tendon Reflex Related Nerve: Sciatic Nerve

Sciatic nerve is the main nerve extended from sacral plexus. So firstly, L_4, L_5 and S1 spinal roots' contribution to sacral plexus was observed and the diameter of each root at the point of intervertebral foramen was measured, and the diameter of each branch contributing to sacral plexus was recorded as well. Then, a picture was taken after the exposure of conus medullaris and cauda equina, as the end of spinal cord cone was marked with a needle, the anterior-posterior position and sequence relationships of $L_4 \sim S_1$ and S_{2-4} anterior and posterior roots were recorded.

2.2.3 Anatomical Study of T_{11-12} Rib Nerves and Spinal Roots Related to Lower Abdominal Wall Reflex

As spinal dura was opened and medullary conus and T_{11}, T_{12} and S_{2-4} spinal roots were exposured, anatomical study about the anterior-posterior position and sequence relationships of T_{11}, T_{12} and S_{2-4} spianl roots were performed.

Stastistical analysis: SPSS 10.0 was used for analysis of ventral spinal roots' horizontal area.

2.3 Results

2.3.1 Anatomical Study of L_{2-4} Spinal Roots and Femoral Nerve (Related to Knee Jerk Reflex)

L_2, L_3 and L_4 spinal roots are the major part contributing to femoral nerve, and there is a hairliking branch of L1 spinal root to join lumbar plexus and femoral nerve which is too small to be measured. L_3 and L_4 spinal roots' contributions (42 % and 39 %) to femoral nerves are great than L_2 spinal roots' (19 %), and there is no statistical difference between L_3 and L_4 spinal roots' (Fig. 2.1).

As femoral is not the only nerve originated from L_2, L_3 and L_4 spinal roots, which also form obturator nerve and L_4 spinal root will send a branch to join the sacral plexus. Figure 2.2 show how much part of L_2, L_3 and L_4 spinal root contributes to femoral nerve: the proportion of L_2 spinal root (86 %) and that of L_3 spinal root (82 %) is more than that od L_4 spinal root (64), $p<0.05$, and there is no statistical difference between that of L_3 and L_4 spinal roots.

Intradural relative position between L2-4 and S2-4spinal roots: It is observed in 39 sides in 20 specimens *that* the level of $S_{2\sim4}$ anterior spinal roots emenate from spinal cord is higher than the level of $L_{2\sim4}$ anterior spinal roots cross meninx, that means and the horizontal area of L_3 or L_4 anterior spinal roots is more than that of $S_{2\sim4}$ anterior spinal roots. Those two levels locate at the same plane is observed in 1 side of 40 sides in 20 specimens.

2.3.2 Anatomical Study of L_4–S_1 Spinal Roots, Sacral Plexus and Sciatic Nerve (Related to Achilles Tendon Reflex)

Sacral plexus are mainly extended from L_4, L_5, S_1 and S_2 spinal roots where L_4, L_5 and S_1 spinal roots' contribution is greater than S_2 spinal root's. Among L_4, L_5 and S_1 spinal roots, L_4 spinal root's contribution (15 %) is less than L_5(46 %) and S_1(39 %) spinal roots' ($p<0.01$), and there is no statistical difference between L_5 and S_1 spinal roots' contribution ($p>0.05$, Figs. 2.3 and 2.4).

Overlapping length between $L_4\sim S_1$ and $S_{2\sim4}$ ventral roots: $S_{2\sim4}$ the level of ventral roots emenate from spinal cord located at L_1 vertebral body level in 28 sides of 20 specimens and it located at L_2 vertebral level in other 12 sides of 20 specimens. The level of $S_{2\sim4}$ ventral spinal roots emanate from spinal cord is higher than the level of $L_4\sim S_1$ spinal roots cross dura, that means, the end part of $L_4\sim S_1$ ventral spinal roots and the beginning part of $S_{2\sim4}$ is overlapping. The overlapping length between $L_4\sim S_1$ and $S_{2\sim4}$ ventral roots is showed in Table 2.1.

2.3.3 Anatomical Study of T_{11-12} Rib Nerves and Spinal Roots Related to Lower Abdominal Wall Reflex

T_{11} and T_{12} spinal roots directly extend into T_{11} and T_{12} rib nerves which relate to lower abdominal wall reflex. There is no statistical difference between the horizontal areas of T_{11}(0.70±0.14 mm^2), T_{12}(0.61±0.18 mm^2) and S_3(0.61±0.07 mm^2), S_4(0.45±0.09 mm^2) anterior roots($p>0.05$), but the horizontal areas of T_{11}, T_{12} anterior roots is

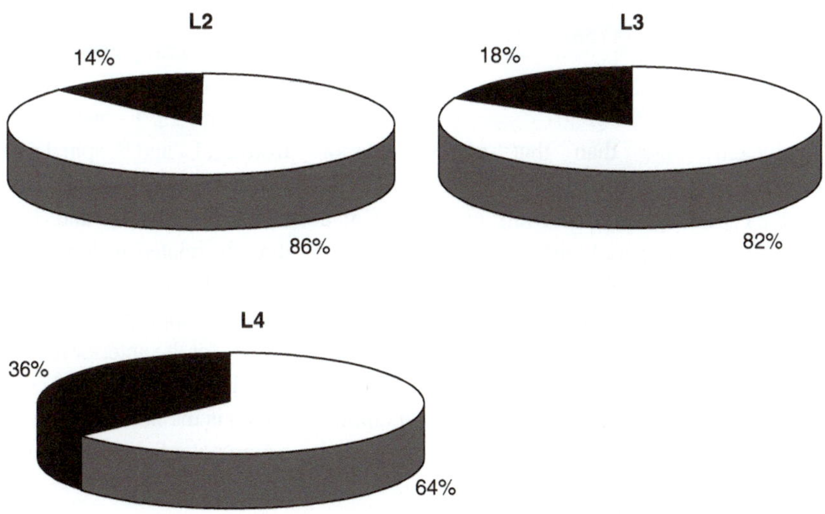

Fig. 2.3 L2, L3 and L4 spinal roots' distribution

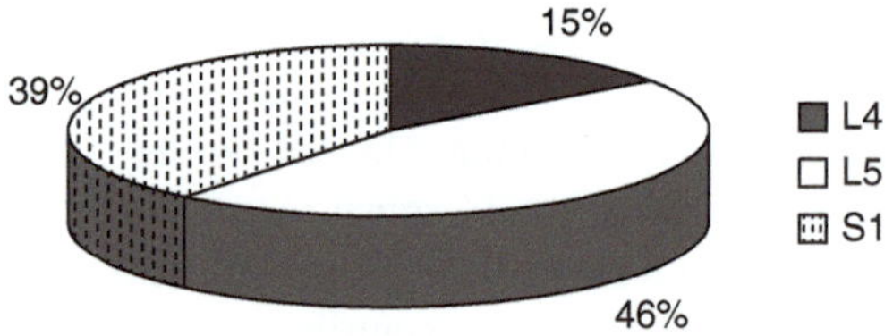

Fig. 2.4 L4, L5 and S1 spinal roots' contribution to sacral plexus

Table 2.1 Overlapping length between $L_4 \sim S_1$ and $S_{2\text{-}4}$ ventral roots $\bar{x} \pm s$ *(Min ~ max)* cm

	L_4	L_5	S_1
S_2	10.3±2.3 (7.1 ~ 13.4)	13.0±2.3 (10.0 ~ 16.7)	15.0±3.1 (12.0 ~ 8.9)
S_3	9.7±2.3 (6.6 ~ 12.3)	12.3±2.2 (9.5 ~ 15.6)	14.4±8.2 (11.5 ~ 17.8)
S_4	8.3±2.4 (4.6 ~ 11.1)	10.9±2.3 (7.5 ~ 13.6)	13.0±2.2 (9.5 ~ 15.8)

Table 2.2 Horizontal area of $L_4 \sim S_4$ ventral spinal roots

Ventral roots' serial	$\bar{x} \pm s$*(min ~ max)*mm^2
L_4	2.19±0.39(2.00 ~ 3.75)
L_5	2.58±0.88(1.77 ~ 3.43)
S_1	2.19±0.42(1.46 ~ 2.75)
S_2	1.12±0.24(0.84 ~ 1.53)
S_3	0.61±0.07(0.47 ~ 0.68)
S_4	0.45±0.09(0.36 ~ 0.59)

Table 2.3 The distance between T_{11}, T_{12} and $S_{2\text{-}4}$ (mm)

		S_2	S_3	S_4
T_{11}	$\bar{x} \pm s$ *(Min ~ max)*	54±25 (31 ~ 90)	60±23 (42 ~ 97)	74±24 (53 ~ 111)
T_{12}	$\bar{x} \pm s$ *(Min ~ max)*	22±20 (3 ~ 56)	28±23 (7 ~ 63)	42±28 (21 ~ 77)

significantly less than that of S_2 anterior roots(1.12±0.24 mm^2), $p < 0.05$ (Table 2.2).

T_{11},T_{12} anterior spinal roots cross dura at a higher plane compared with that $S_{2\text{-}4}$ anterior spinal roots emenate from spinal cord, so there is a space between the end points of $T_{11,\ 12}$ anterior roots and beginnings of $S_{2\text{-}4}$, the detail is show in Table 2.3.

Intradural relative position between spinal roots: All specimens' medullary conus are ended at the plane of $L_1 \sim L_2$ vertebral body level where $S_{2\text{-}5}$ ventral and dorsal roots just send out from spinal cord and can be distinguished from each other according to the denticulate ligament and vessel between them, but it's still difficult to judge the series of S_2, S_3 and S_4 just based on its level coming from spinal cord. The size of L_4, L_5 and S_1 spinal roots are big enough to distinguish ventral and dorsal roots, also when one spinal root was confirmed others can be distinguished successively because their thickness make it easy.

Horizontal area of L_4, L_5, S_1 and $S_{2\text{-}4}$ ventral roots

Every ventral root is composed of 1 ~ 3 bundles. Among L_4(2.19±0.39), L_5(2.58±0.58) and S_1(2.19±0.42) mm^2, L_5 is the Largest one and L_4, L_5, S_1 ventral roots are significantly larger than S_2(1.12±0.24), S_3(0.61±0.07) or S_4(0.45±0.09) mm^2 ventral roots, $p < 0.01$ (Fig. 2.2).

2.4 Discussion

2.4.1 Select the Suitable Spinal Root as Dynamic Nerve

Due to the embryonic sarcomere split, migration and re-combination, a certain nerve can contain kinds of nerve fibers from different spinal roots and nerve fibers from a single root can be distributed to many peripheral nerves [5]. Femoral nerve, as the knee reflex related nerve, is extended from lumbar plexus orginated from L_2, L_3 and L4 spinal roots, while nerve fibers from L_2, L_3 and L_4 spinal roots will extend as femoral nerve, obturator nerve and parts of sacral plexus [6]. Sacral plexus and sciatic nerve, as the Achilles tendon reflex nerve, are orginated from L_4, L_5 and S_1 nerve, while nerve fibers from L_4, L_5 and S_1 spinal roots will extend as sciatic nerve, superior gluteal nerve, inferior gluteal nerve etc. [7]. Only T_{11} and T_{12} intercostal nerves are directly related to lower abdominal muscle and a stable skin area [8]. Therefore, it's important to select the appropriate spinal nerve root as "dynamic nerve" for the anterior root cross anastomosis and to establish "tendon-spinal cord-bladder" nerve reflex, which is the nerve linking basis for spreading more action potential to the bladder [9] (Fig. 2.5).

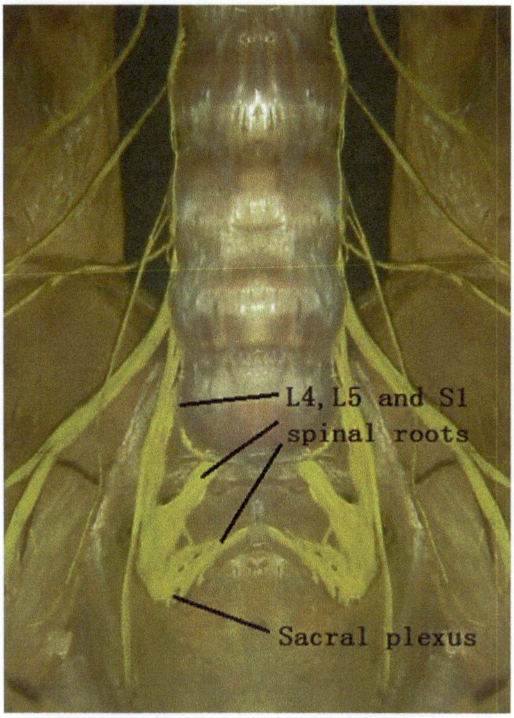

Fig. 2.5 Composition of sacral plexus composed of L4, L5 and S1 spinal roots mostly

The afferent and efferent nerve fibers of knee tendon reflex are located in femoral nerve, which is originated and extended from L_2, L_3 and L_4 spinal roots, and L_3 and L_4 spinal roots contribute more to it than L_2 spinal root does [6]. On the one hand, from point of this view, L_3 and L_4 spinal roots should be selected as dynamic nerve to ensure more action potential of knee reflex can be translate through the certain spinal root. On the other hand, as to a single spinal root, how much part of nerve fibers will contribute to femoral nerves is also important. This anatomical study shows that L_2 and L_3 spinal roots' are more than that of L_4 spinal root, so from point of this view, L3 spinal root is better than L_4 spinal root. L_3 is the best as dynamic nerve and L_3–S_3 or S_4 anterior spinal roots cross anastomosis should be taken during the operation of bladder functional reestablishment by the use of knee jerk reflex.

Sciatic nerve is the largest one orgined from Sacral plexus which is composed of L_4, L_5, S_1 and S_2 spinal roots. As sciatic nerve extends as tibial nerve and common peroneal nerve, the former contains afferent and efferent nerve fibers of Achilles tendon reflex [2, 7]. Because L_4, L_5, S_1, S_2 and S_3 spinal roots will firstly mix and knit as sacral plexus before extending as sciatic nerve and tibial nerve, it's difficult to calculate every spinal root's contribution to tibial nerve just through anatomy study. The proper method is to stimulate each spinal root and observe constriction of gastrocnemius muscle and movement of ankle joint, and select the one which triggers off the most forcefully constriction of gastrocnemius muscle. At the same time, this study show that each spinal root of L_4, L_5 and S_1 is composed of 2~3 branches, so the same method can be used to select the proper branch as dynamical nerve to make anastomosis with S_3 or S_4 ventral root so as to establish "Achilles tendon-spinal cord-bladder" reflex and achieve the best effect (Fig. 2.6).

Fig. 2.6 Origin of sciatic nerve-a posterior view

Far extremity of this cross anastomosis should select the root which has predominant nerve fibers to bladder, so that when "tendon-spinal cord-bladder" were established, bladder can gain more nerve control. According to our prior study, S3 or S4 ventral root should be selected as far extremity of the cross anastomosis [10, 11].

2.4.2 Select Ideal Place for Anastomosis Operation

As S2 ~ 4 roots exit from spinal cord at the level of $L_{1\sim2}$ vertebra body level, cross anastomosis of a dynamic spinal root to S_3 or S_4 can be finished at the segment of this level to L_5 or S_1 exit from dura. When cross anastomosis level near to $L_5 \sim S_1$ vertebral body horizontal plane, it's easy to distinguish L_3 or L_5 ventral root according to it's position at L_3 or L_5 intervertebral foramen, so that, as operation is done at this region, to distinguish L_3 or L_5 as dynamical nerve is possible and easy. But $S_{2\sim5}$ ventral roots always enwrapped in one bundle at this region, although it's reported that to distinguish ventral, dorsal roots and their sequence can be done according to their relatively position [5, 12, 13], in fact, $S_{2\sim5}$ roots are thin and ventral roots are in deep position, all those make it difficult to distinguish ventral and dorsal roots or identify their sequence. Even electric stimulation method can work because ventral roots can arise constriction of bladder whereas dorsal root doesn't. As a result, it's not an ideal operation position.

If operation is done at the region of terminal cone level, on the one hand, ventral roots and dorsal roots just exit from spinal cord and apart from each other by denticulate ligament and vessel on the surface of spinal cord, so to distinguish ventral roots and dorsal roots is possible and easy. At the same time, when L_1 root was affirmed according to it get across L_1–L_2 intervertebral foramen, L_1 ventral root can be find just move L_1 dorsal root away and L_2, L_3, L_4, L_5, S_1 ventral roots can be distinguished successively because they are large and arrange from lateral to medial way. Dynamical nerve can be correctly selected at this level. On the other hand, thin as S2 ~ 4 spinal roots are, but they can be identified from each other at the surface of spinal cord, combined with electric stimulation and bladder pressure measure, the proper far extremity of cross anastomosis can be correctly determined.

Therefore, the ideal place for cross anastomosis operation is at terminal cone level, which can be confirmed by CT scan before operation.

2.4.3 Indication of the Operation for Bladder Functional Reconstruction with Lower Abdominal Reflex, Knee Tendon Reflex or Achilles Tendon Reflex

As CNS of lower abdominal reflex locates in T_{11} and T_{12} spinal cord, knee tendon reflex in L_3, and achilles tendon reflexes in the L_5 and S_1 cord level. Which one should be selected to establish an artificial reflex to recover bladder function? It depends on which reflex was interrupted after spinal cord injury. For example, When spinal cord injure level is lower, especially when CNS of jerk reflexes and lower abdominal reflexes is injured, achilles tendon reflex can be select to establish "achilles tendon-CNS-bladder" for bladder functional regain.

Conclusion

1. During the operation of bladder functional reestablishment by the use of lower abdominal reflexes as the T_{11}/T_{12}–S_3/S_4 anterior spinal roots cross anastomosis are taken, the horizontal area of T_{11}, T_{12} anterior spinal roots are suitable for cross anastomosis. The nerve grafting most be taken as the cross anastomosis of T_{11}, T_{12} and $S_{2\sim4}$ anterior spinal roots because of the distance between them (Fig. 2.7).
2. L_3–S_3 or L3–S_4 anterior spinal roots cross anastomosis should be taken during the operation of bladder functional reconstruction by the use of knee jerk reflex.
3. L_5/S_1–S_3 or L_5/S_1–S_4 anterior spinal roots cross anastomosis should be taken during the operation of bladder functional reconstruction by the use of Achilles tendon reflex.
4. When artifical reflex were performed for bladder functional reestablishment, cross-anastomosis of spinal anterior roots should be done at the level of terminal cone, so as to easily distinguish the anterior spinal roots from posterior roots and identify their sequence for those three kinds operations.

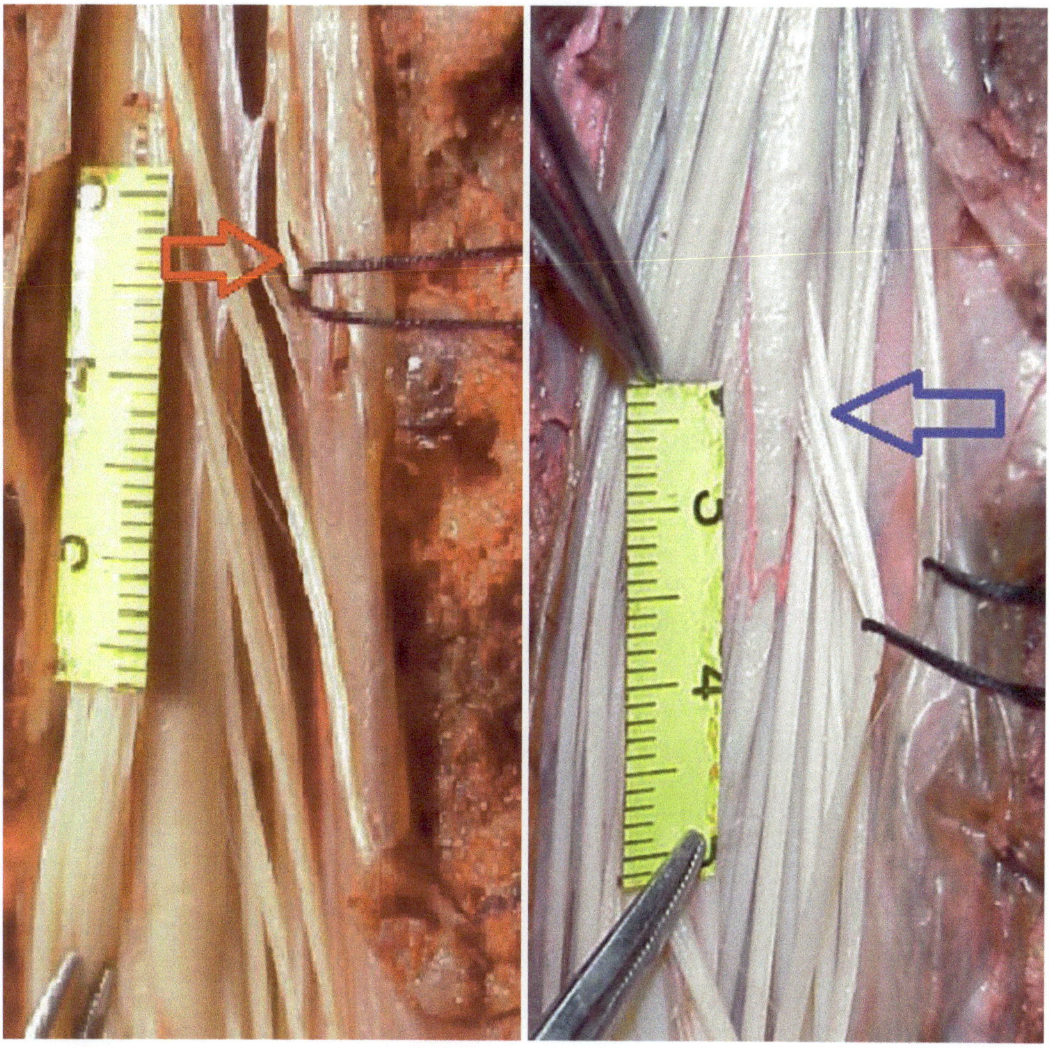

Fig. 2.7 T11, T12 anterior roots diameter is the same as S3, S4 anterior roots and less than S2 anterior roots

References

1. Tanagho EA, Schmidt RA. Electrical stimulation in the clinical management of the neurogenic bladder. J Urol. 1989;140:1331–9.
2. Hou CL, Zhang SM, Zong HB. Emiction function reconstruction after spinal cord injure. Mod Rehab. 2000;4:815–7.
3. Xu RS, Hou CL, Zhang SM, et al. An animal model establishment of bladder functional reconstruction with artificial bladder reflex arc. Chin J Spin Spin Cord. 2002;12:102–4.
4. Zheng XY, Hou CL, Zhong HB, et al. Reconstructed bladder innervation below the level of spinal cord injury: the knee-tendon to bladder artificial reflex arc. J Spinal Cord Med. 2009;32:79–85.
5. Wall EJ, Cohen MS, Massie JB, et al. Cauda equina anatomy I: intrathecal nerve root organization. Spine. 1990;15:1244–7.
6. Zhong SZ. Clinical anatomy. Beijing: People's Health Press; 1998. p. 339–410.
7. Zhang CY. Human anatomy. 2nd ed. Beijing: People's Health Press; 1998. p. 1680–2.
8. Xu RS, Hou CL, Zhang SM, et al. Comparison of different dynamic anterior roots on bladder functional reconstruction. Chin J Orth Surg. 2002;22: 226–8.
9. Xu RS, Hou CL, Zhang SM, et al. Applied anatomical and clinical study on bladder functional reconstruction

with 'Knee-Spinal Cord-Bladder' after spinal cord injury. Chin J Orth Surg. 2002;10:869–71.
10. Xu RS, Hou CL, Zhang SM, et al. Applied anatomical study for the operation of bladder functional reconstruction with lower abdominal reflexes. Chin J Clin Anat. 2002;20:275–6.
11. Xu P, Xu DC, Zhang WB, et al. Clinical anatomy of limited selective posterior rhizotomy at the caudal of medullary cone. Chin J Clin Anat. 1999;17:1–3.
12. Zhang SM, Hou CL. Treatment of spasmus bladder after spinal cord injury with denervation. Mod Rehab. 2000;4:812–5.
13. Zong HB. A study about bladder reinnervation. Mod Rehab. 2000;4:818–9.

3 Classification of and Treatment Principles for Bladder Dysfunction Caused by Spinal Cord Injury

Mingxuan Liu and Chunlin Hou

3.1 Introduction

Bladder dysfunction caused by spinal cord injury is one of the most difficult clinical problems not yet solved. It often causes severe urinary incontinence, urinary retention and urinary tract infection, or even renal failure leading to death [1, 2]. Exploration of ways to solve this problem is very important to the improvement of paraplegic patients' quality of life and reduction of their mortality rate. Urodynamic examination has an exact value in the diagnosis, classification, treatment and prognosis of lower urinary tract dysfunction of patients with spinal cord injury. It can classify bladders, thereby providing timely objective indicators on which to base the diagnosis and treatment of patients with lower urinary tract dysfunction. We performed a systematic examination of lower urinary tract function of 136 patients with spinal cord injury, established a relatively detailed and systematic classification and set treatment principles correspondingly.

M. Liu • C. Hou, M.D. (✉)
Department of Orthopedic Surgery,
Changzheng Hospital,
The Second Military Medical University,
Fengyang Road 415, Shanghai 200003,
People's Republic of China
e-mail: 1988001@sina.com; chunlin_hou@163.com

3.2 Materials and Methods

3.2.1 Clinical Data

Among the 36 patients, 28 were male and 8 female, ranging in age from 17 to 55 years. The course of the disease ranged from 2 months to 22 years. Spinal cord injuries were all caused by trauma: 6 cases were incomplete damage and 30 were complete damage; 21 cases were suprasacral cord injury and 15 were sacral cord injury. All sacral cord injuries were confirmed by CT scan, magnetic resonance examination or surgical observation. According to the degree and site of their injury, the 36 patients were divided into four groups: A, B, C and D.

Group A: 6 cases, all incomplete spinal cord injury, 3 suprasacral cord injury cases and 3 sacral cord injury cases. Reflex bladder was established in all the patients. Their urination function was close to normal. The patients had partial feeling during urination, the starting phase of urination could be consciously controlled and there was very little or no residual urine. There was a history of urinary tract infections in 1 patient. None of the cases had renal impairment.

Group B: 5 cases, all complete thoracic spinal cord injury. Patients established automatic reflex bladder (i.e., stimulating a specific site could cause urination). Their urination function was close to normal. Patients had no feeling during urination, the starting phase of urination could not be consciously controlled and there was

C. Hou (ed.), *Functional Bladder Reconstruction Following Spinal Cord Injury via Neural Approaches*,
DOI 10.1007/978-94-007-7766-8_3, © Springer Science+Business Media Dordrecht 2014

very little residual urine. There was a history of urinary tract infections in 1 patient. None of the cases had renal impairment.

Group C: 13 cases, all complete cervical and upper thoracic spinal cord injury. Patients had different degrees of bladder spasm, and those with a serious degree of spasm urinated frequently. There was bladder contracture in 3 patients with a longer course of the disease. Some of the patients had urinary retention. Autonomic dysreflexia was found to different degrees in 13 patients, renal impairment in 5 patients and a history of urinary tract infection in 6 patients

Group D: 12 cases, all complete sacral spinal injury. The patients' clinical manifestations were all atonic bladders. Some patients had urinary incontinence. Patients could not urinate by themselves and needed to rely on abdominal oppression or a catheter to complete voiding. There were large amounts of residual urine. There was renal impairment in 1 patient and a history of urinary tract infection in 12 patients.

3.2.2 Test Method

Using a four-lead urodynamic pressure capsule, examine indicators including change in bladder volume pressure, detrusor function, myoelectric activity of the urethral sphincter and urethral pressure. Perform residual bacteria cultivation, renal function test and bladder intravenous pyelography (IVP) in every case.

3.3 Results

Each group was tested respectively and the test results were (Table 3.1):

1. Bladder volume pressure curve: the results of Groups A and B are close to normal and there were very few reflected waves. The slope of Group C increased, and the curve shifted left. There were many reflected waves. Group D had very flat curve that shifted right, and there was no reflected wave.
2. Detrusor reflection: Groups A and B had hyperreflexia, Group C had obvious hyperreflexia and Group D had no hyperreflexia.
3. Bladder compliance: Groups A and B were close to normal. Group C had low compliance, while Group D had high compliance.
4. Electromyography of the detrusor and sphincter: The detrusor and sphincter harmonized with each other in Groups A and B while they were inharmonious in Group C. The detrusor and sphincter were unrelated with each other in Group D.

The test results showed that urodynamic performance varies according to the different site and degree of spinal cord injury. Different urodynamic performance leads to different bladder function and complications. In this regard, the urodynamic classification of the bladder is very important to the prevention and treatment of complications, as well as the improvement and recovery of bladder function.

3.4 Discussion

Clinical manifestation of the bladder after spinal cord injury is complicated, and there are many ways of classification. Turner-Warwick is a frequently used method that classifies patients' bladder into detrusor hyperreflexia and detrusor areflexia.

Table 3.1 Urodynamic and other examination results of the 36 patients

N		Bladder capacity (ml)	Bladder pressure (KPa)	Urethral pressure (KPa)	Residual volume (ml)	Urine culture (+)	Urine return	Urinary tract infection	Renal damage
A	6	300±100	7±2	6±1	40±10	0	0	1 (16 %)	0
B	5	300±100	7±2	6±1.5	50±10	3	0	1 (20 %)	0
C	13	150±100	9±3	8±1.5	100±30	25	4	7 (55 %)	5 (38 %)
D	12	500±200	4±1	4±1.5	200±100	31	0	10 (83 %)	1 (8.3 %)

After systematic urodynamic examination, we found that bladder function is determined by three factors: (1) detrusor reflex contraction function; (2) urethral sphincter contraction function; (3) cooperation between the detrusor and sphincter. Factor 3 is especially vital to the maintenance of normal urination. Therefore, the author divided bladders after spinal cord injury into the following three classes based on Turner-Warwick:

1. **Bladder with detrusor hyperreflexia and coordinating sphincter;**
2. **Bladder with detrusor hyperreflexia and disordered sphincter;**
3. **Bladder with detrusor areflexia.**

In the first class, there is no sphincter activity during detrusor reflex contraction. There is no resistance in the urethra, and the bladder is able void smoothly. Cases in Group A and B all belong to this classification. The urodynamic performance of patients in these two groups is quite similar. Patients have all established reflex bladder. Their bladder volume pressure curve, bladder capacity and urethral pressure are close to normal. Therefore, patients in this class have close-to-normal urination function and very little residual urine. In this regard, the urinary infection rate is very low. Renal impairment is also rare.

According to the test results of these two groups together with clinical manifestation, we found that: (1) incomplete spinal cord injury leads to different degrees of detrusor hyperreflexia; probably the spinal cord is not completely damaged and can still maintain a neural arc. (2) If the case is suprasacral cord injury, the lower the injury site, the greater probability that reflex bladder can be established and the better the urination function. The 5 patients in Group B all suffered from thoracic spinal cord injury. (3) If the bladder of the first class is identified in the early stage of spinal shock recovery, there is great potential for patients' spinal function to be improved. Before the limb function of the 6 patients in Group A (incomplete spinal cord injury) recovered, urodynamic examination of all of them indicated myoelectric coordination of the detrusor and sphincter. The author thinks that this kind of bladder should be trained as early as possible so as to restore its urination function as soon as possible.

Treatment principles for such patients:

1. Further enhance the training of autonomic bladder to restore physiological micturition function.
2. Prevent urinary system infection and renal function damage.

The second class of bladder is characterized by a sudden increase in sphincter myoelectric activity when there is detrusor reflex contraction. At this time, urethral resistance increases significantly and the urine in the bladder cannot be voided smoothly. The bladders of the 13 patients in Group C (cervical and upper thoracic spinal cord injury) belonged to this class. The volume pressure curve visibly shifted left, and there were several reflected waves. Both bladder and urethra pressure rose. The most common clinical feature is bladder spasm and frequent urination. Some patients may suffer urinary retention because of the increase of resistance in the urethral meatus. High bladder pressure may also cause autonomic dysreflexia and even convulsions and cerebral hemorrhage. Patients with a long course of the disease may suffer from bladder contracture. The urinary system infection rate of this group is 55 %, and the rate of renal damage is 38 %, which is far higher than in other groups.

Correlated dates indicate that renal function impairment, renal failure and renal amyloidosis are the major cause of death of paraplegic patients at later period [3]. The reason for renal function impairment is much disputed. Based on the examination results, the author thinks the primary cause is the high bladder pressure resulting from a synergistic function disorder of the detrusor and sphincter, which leads to urine reflux and then high kidney pressure. Frequent and long-term high pressure in the kidney leads to renal impairment. Failure in timely treatment of residual urine also leads to urine reflux, which is the secondary cause of renal impairment. Urinary infection, together with the above two factors, will aggravate renal impairment. Angiography results of this group showed that 4 patients with urine reflux all suffer from renal function impairment, which indicates that urine reflux is closely related to renal impairment. Patients with such bladder usually suffer the worst complications, and the death rate is also very high.

Treatment principles for such patients:

1. Adopt a positive treatment method. Reduce urethral resistance through medicine or surgery.
2. Relieve detrusor spasm through surgery (i.e., selective sacral rhizotomy) or medicine to restore bladder equilibrium function so as to avoid the above complications.
3. Actively prevent and treat urinary system infection and renal impairment.
4. Perform positive training of autonomic bladder after the bladder equilibrium function is restored.

The third class of bladder has the following features: there is no detrusor contraction. The bladder volume pressure curve is flat and shifts rightward. Urethral pressure is low, and bladder capacity is large. Myoelectric activity of the sphincter is irrelevant to detrusor conditions. The 12 patients in Group D all had bladder of this class. Some of the patients suffered from urinary incontinence. Because of poor voiding ability, there is usually large amount of residual urine. The urinary system infection rate of this group is 83 %, and the rate of renal damage is 8.3 %. The infection rate is far higher than in other groups.

The test results indicated that residual urine is closely relevant to urinary infection. The author thinks that the large amount of residual urine causes the urine in the bladder bottom to remain in a relatively stable condition for a long time. It is difficult to void bladder bottom urine. The residue, epicyte and bacteria are largely deposited. Infection will occur once the time is ripe. When we examined the patients with a long course of the disease and a large amount of residual urine, we found that the residual urine we drained from the bladder was a thick liquid and was proved to contain a large amount of bacteria. Another factor that leads to infection is the lack of muscle tone of the detrusor. A large amount of urine will make the bladder wall overinflate. Research indicates that such a condition could reduce the antibacterial capacity of the bladder and therefore provide an environment conducive to bacterial invasion [4, 5].

Treatment principles for such patients:

1. Improve bladder function through surgery or medicine to eliminate residual urine.
2. Relieve urinary retention and eliminate residual urine through amicrobic or apinoid catheterization.

The above two principles are currently the most effective way to solve urinary system infection caused by the third class of bladder.

References

1. Wurster RD, Randall WC. Cardiovascular responses to bladder distension in patients with spinal transection. Am J Physiol. 1975;228:1288–92.
2. Kurmick NB. Autonomic hyperreflexia and its control in patients with spinal cord lesions. Ann Intern Med. 1956;44:678–86.
3. Hackler RH. A 25-year prospective mortality study in the spinal cord injured patient: comparison with the long-term living paraplegic. J Urol. 1977;117:486–8.
4. Ross JC, Gibbon NOK, Sunder GS. Division of the external urethral sphincter in the neuropathic bladder: a twenty years' review. Br J Urol. 1976;48:649–56.
5. Lapides J, Diokno AC, Silber SJ, Lowe BS. Clean intermittent self-catheterization in the treatment of urinary tract disease. J Urol. 1972;107:458–61.

4 Pathological Changes in the Detrusor After Spinal Cord Injury

Xianyou Zheng and Chunlin Hou

4.1 Introduction

Among the many types of spinal fractures reported, thoracolumbar fracture is the most common; it is characterized by medullary cone injury, functional impairment of the sensory and motor nerves and atonic bladder [1–3]. Atonic bladder leads to severe urinary retention, refractory urinary infections, and even chronic renal failure, making it the main cause of death from medullary injury [4, 5]. Therefore, it is important to restore self-controlled urination in these patients, in order to decrease the mortality rate and improve their quality of life.

Functional restoration of atonic bladder involves establishment of an intradural autogeneic neural anastomosis with the normal somato-reflex, above the paraplegic level, for example, the abdominal reflex [6–8]. This technique was used to successfully establish a skin-spinal cord-bladder reflex path in rats: the anastomosis was made between the central end of the ventral root of L4 and the peripheral end of the ventral root of L6, with the dorsal root of L4 intact. We believe that establishment of an artificial bladder reflex with a somatic nerve-to-autonomic nerve anastomosis has far-reaching influences on atonic bladder reconstruction. In our recent studies, we successfully established an abdominal reflex-spinal cord-bladder reflex path to restore controlled urination in rats [9]. Moreover, we used this method to successfully treat a female patient with medullary cone injury caused by an L1 fracture [10]. Four months after the injury, we established a new technique wherein we created an anastomosis between the ventral root of T11 and the ventral root of S2 with sural nerve transplantation. Self-controlled urination was successfully restored in the postoperative 55 months. Furthermore, it was confirmed using urine dynamics monitoring that the urination was controlled completely by constriction of the detrusor muscles. We have therefore verified the reliability of the abdominal reflex-spinal cord-bladder reflex path in restoring controlled urination both experimentally and clinically.

Atonic bladder is a manifestation of lower motor neuron paralysis; it can therefore result in degeneration of the target organ because of denervation and loss of nutrition, which eventually lead to decrease in muscular tone and disappearance of the physiological reflex [11, 12]. Therefore, we think that nerve repair procedures should be performed before procedures are conducted on the detrusor muscle (DM) and its neuromuscular junction (NMJ). In order to study this further, it is first necessary to understand clearly the time-course changes that occur in the DM and its NMJ after medullary cone injury.

X. Zheng
Department of Orthopedic Surgery, Shanghai Sixth People's Hospital, Shanghai Jiaotong University, Shanghai, People's Republic of China
e-mail: zhengxianyou@126.com

C. Hou, M.D. (✉)
Department of Orthopedic Surgery, Changzheng Hospital, The Second Military Medical University, Fengyang Road 415, Shanghai 200003, People's Republic of China
e-mail: chunlin_hou@163.com

C. Hou (ed.), *Functional Bladder Reconstruction Following Spinal Cord Injury via Neural Approaches*, DOI 10.1007/978-94-007-7766-8_4, © Springer Science+Business Media Dordrecht 2014

In this study, we present a rat model for atonic bladder and observe morphological changes in the DM and its NMJ. Moreover, we have investigated time-course degeneration in the DM and its NMJ after medullary cone injury, as this information could help determine the optimal operative time for restoring self-controlled urination in patients with atonic bladder.

4.2 Materials and Methods

4.2.1 Animals and Treatments

All procedures were approved by our institutional review committee and followed the national guidelines for animal care and experimentation. We randomly assigned 56 adult female 7-week-old Sprague–Dawley (SD) rats to seven groups (n = 8 for each group). All rats were individually housed in cages. The seven groups comprised the normal control group and the 4, 6, 8, 10, 12 and 16 weeks after medullary cone injury groups; they were named in order as A–G. The SD rats of groups B–G were considered as atonic bladder models. The rats were weighed and anesthetized with an intraperitoneal injection of 1 % sodium pentobarbital (40 mg/kg), and were then immobilized with abduction of the extremities in the prone position. After degermation, medullary cone injury was induced at the level of the L4/5 vertebral body in a sharp and transsectional manner, with the use of an operating microscope. After the operation, the wound was disinfected regularly until it healed. The rats underwent abdominal pressing 2–3 times per day for urination. They were observed for ulcerations in the extremities, infections of the wounds, self-eating of the limbs and survival rates. After procurement of the DMs, the following studies were conducted.

4.2.2 Histological Studies of the DM

4.2.2.1 Sectional Areas of Muscle Fibers

The DMs were first fixed for 48 h in 10 % formalin. After dehydrolysis and paraffin embedding, transversal slices of 5 μm thickness were taken from the middle of the DM. Then, hematoxylin and eosin (HE) staining was carried out using the FW4000 digital imaging workstation (Leica Inc. Germany). The average value of the sectional areas of muscle fibers was finally calculated.

4.2.2.2 Ultrastructures of the DM and Its NMJ

The DMs were fixed with 4 % glutaraldehyde and osmic acid individually; they were then were dehydrated with pyroracemic acid, embedded with EPON812, sliced with the LKBNOVA ultramicrotome (Bromma Inc., Sweden), and dyed with lead and uranium individually. Finally, the ultrastructures of the DM and its NMJ were observed under the Philips-CM120 transmission electron microscope (Philips Inc., Dutch).

4.2.3 Analysis of Fibrosis in the DMs

The muscles were subject to Masson trichrome staining: (1) The DM was fixed in an alcohol-glacial acetic acid-formaldehyde solution (AAF), dehydrated, embedded with paraffin, and sliced to 2 μm thickness. Finally, the slice was deparaffinized. (2) The slice was fixed with Bouin solution for 30 min and then flushed with lotic water. (3) The nucleus was dyed with hematoxylin for 10 min and was then flushed with lotic water. (4) The section was dyed with Ponceau red 2R · acid fuchsin · rubrum scarlatinum for 10–20 min and then flushed with distilled water. (5) The slice was dyed with 1 % phosphomolybdic acid for 3–5 min, and then dyed with 2 % aniline blue for 5–10 min. (6) After differentiation with a 0.5 % acetic acid solution, the slice was dehydrated and mounted with neutral gum. The collagen fibers were stained blue; muscular fibers, red; endochylema, red; nucleus, rufous; and red blood cells, salmon pink. Masson trichrome staining was performed in order to calculate the percentage of smooth muscles and connective tissue, using the FW4000 digital imaging workstation.

4.2.4 Immunohistochemical Procedures

α-Smooth muscle antibody (α-SMA) is a microfilament protein that is required for contraction of muscles; it is therefore an indicator of muscle contractability. α-SMA was stained using an automatic immunohistochemical staining system (Ventana Medical Systems Inc., USA). The primary antibody was anti-α-SMA (Santa Cruse Inc., USA), and the secondary antibody was provided by Ventana Medical Systems Inc. The integral optical density (IOD) of the α-SMA was analyzed using the FW4000 digital imaging workstation. Ten fields (10×10) were chosen in each section to measure the mean IOD of α-SMA.

4.2.5 Statistical Analysis

Means and standard deviations were calculated from individual values using standard procedures. One-way analysis of variance (ANOVA) was used to determine significant differences between the groups, and the Student–Newman–Keuls (SNK) test was used for multiple comparisons. Differences were considered significant at $P<0.05$.

4.3 Results

All of the rats survived; no infections or self-eating of the limbs was observed. One rat in each of the groups E, F and G had ulcerations in the extremities. After disinfection and bandaging of the affected extremities, no further deterioration occurred (Table 4.1).

4.3.1 Sectional Areas of DM Fibers

In group A, the DM fibers were in fusiform shape and funiculose; the cellular nucleus had consistent contours; and no infiltration of connective tissue could be detected in the muscular bundles. However, from groups B–G, the muscular fibers were gradually more crescent-shaped or irregular; they showed misalignment and disarray; and marked infiltration of connective tissue was observed in the muscle bundles (Fig. 4.1). Gradually progressive atrophy of the DM over time was found in groups B–G (Fig. 4.2). The sectional areas of the DM fibers in groups A–G were 7265.86±114.36,5156.31±310.42,5398.47±453.08, 4146.49±284.06, 3067.40±130.91, 2123.56±235.66, and 1675.98±322.09 μm^2, respectively. The sectional areas of muscle fibers in groups E, F, and G were significantly lower ($P<0.05$) than that in control group A, but there was no significant difference between groups B, C, and D, and group A.

4.3.2 Ultrastructure of the DM

In group A, DM cells were well-arranged and well-distributed with consistent contours and size, and normal intermediate junctions were observed, with the distance between muscular

Table 4.1 Detrusor muscle changes in rats after medullary cone injury (n=8, X±s)

Group	Bladder weight (g)	Sectional area of muscle fibers (μm^2)	Percentage of connective tissue (%)	α-SMA activity (IOD)
A	0.138±0.019	7,265.86±114.36	12.08±1.45	0.96±0.03
B	0.286±0.025	5,156.31±310.42	18.47±1.32	0.85±0.10
C	0.362±0.018	5,398.47±453.08	23.24±2.76	0.74±0.06
D	0.412±0.082	4,146.49±284.06	26.29±1.88	0.69±0.09
E	0.496±0.022[a,b]	3,067.40±130.91[a,b]	35.67±1.12[a,b]	0.44±0.08[a]
F	0.589±0.023[a]	2,123.56±235.66[a]	47.22±2.38[a,b]	0.39±0.10[a]
G	0.721±0.020[a]	1,675.98±322.09[a]	55.56±3.12[a]	0.36±0.07[a]

[a]Compared with Group A, $P<0.05$
[b]Compared with Group G, $P<0.05$

cells being much smaller than that in abnormal specimens. A few collagen fibers and elastic fibers could be seen in the matrix. Organelles such as the mitochondria and endocytoplasmic reticulum, myofilaments and dense bodies were well-organized in the smooth muscle cells. By contrast, in groups B–G, the atonic bladders showed abnormalities in the ultrastructure of detrusor cells that were gradually aggravated over time, such as inconsistent contours, misalignment and disarray; wide separation between muscular cells; abundant collagen fibers and irregular dense structures between individual cells; obviously widened rough endoplasmic reticulum (ER) and mitochondrial edema (Fig. 4.3).

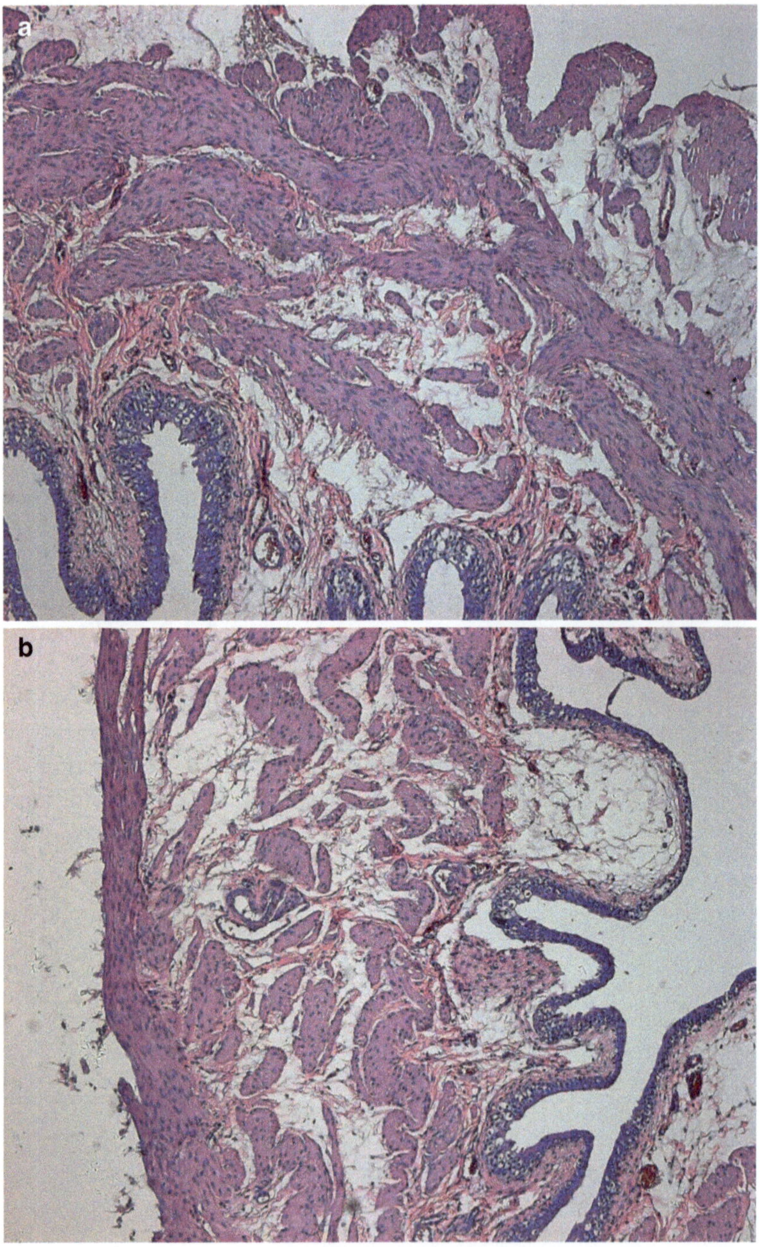

Fig. 4.1 Transverse sections of the detrusor muscle in rats of the normal control (**a**), and 6 weeks (**b**), 10 weeks (**c**) and 16 weeks (**d**) after medullary cone injury groups. Hematoxylin and eosin staining (×100)

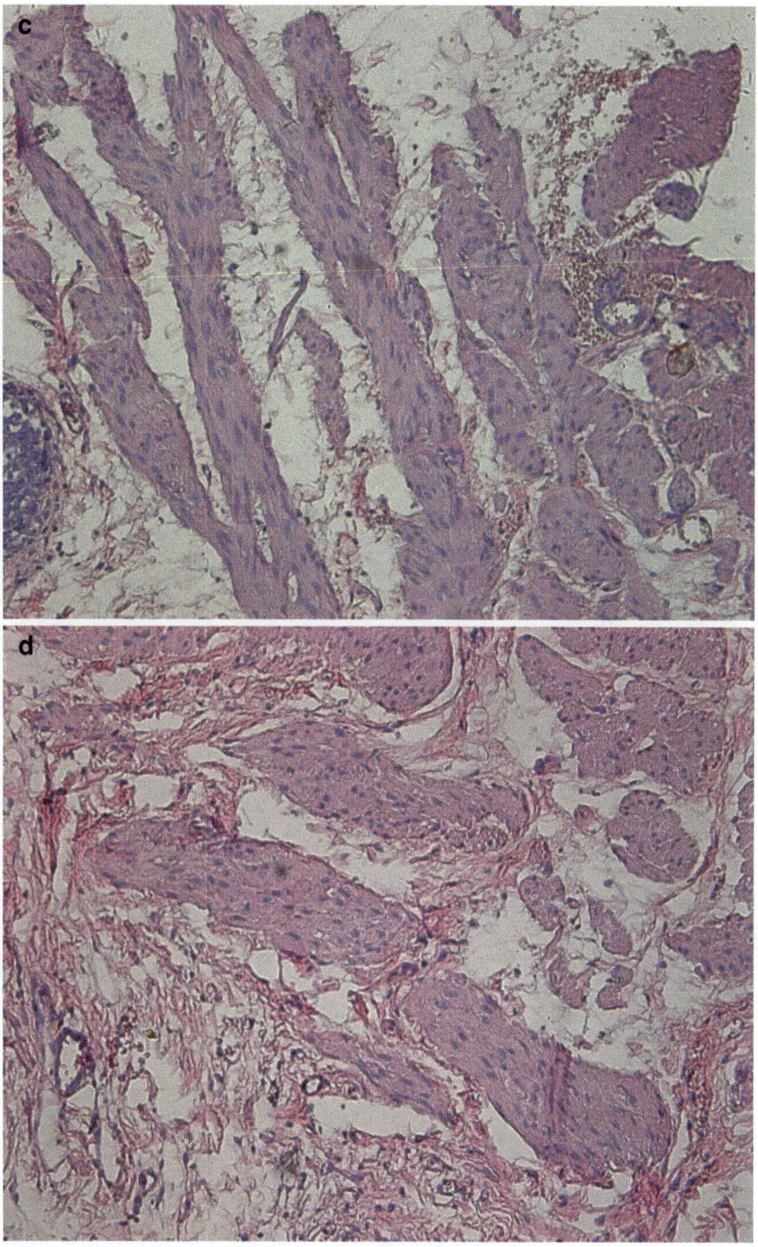

Fig. 4.1 (continued)

4.3.3 Ultrastructure of the NMJ

In group A, the normal NMJ contained a large amount of organelles, such as mitochondria and ER, and abundant synaptic vesicles in the sympathetic nerve endings. The NMJ in groups B, C and D showed a lower amount of mitochondria and synaptic vesicles. Conspicuous degeneration of the NMJ was observed in group E; it was manifested as derangement or disappearance of the reductus, and obvious reduction in the amount of synaptic vesicles, mitochondria and ER in the sympathetic nerve endings. Further deterioration and obvious reduction or disappearance of synaptic vesicles and mitochondria, as well as degenerative corpuscles, was noted in group F. Finally, in group G, severe deterioration of the NMJ accompanied by absence of synaptic vesicles and

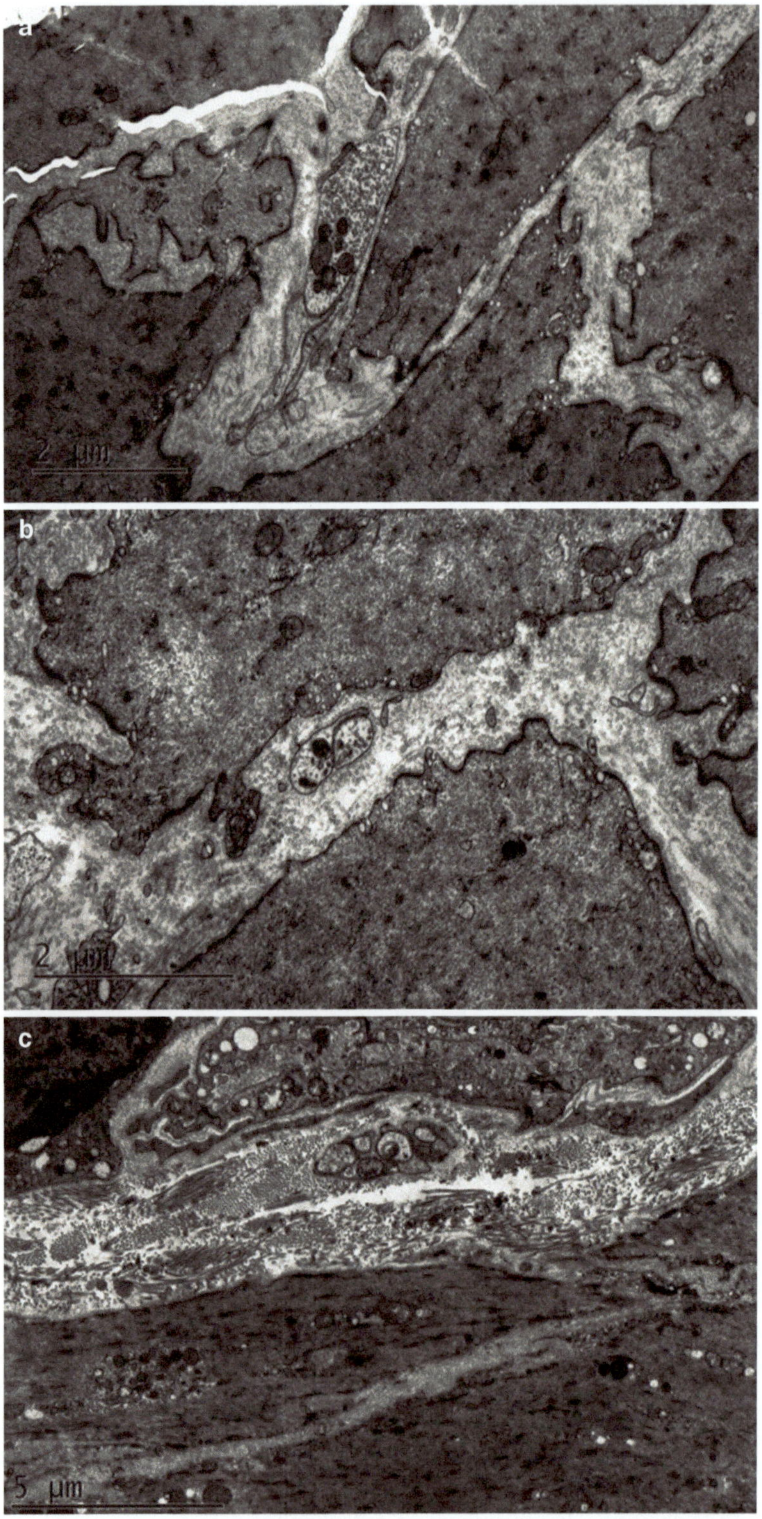

Fig. 4.2 Ultrastructures of the detrusor muscle and its neuromuscular junction observed under a transmission electron microscope in rats of the normal control (**a**), and 6 weeks (**b**), 10 weeks (**c**) and 16 weeks (**d**) after medullary cone injury groups. *1* detrusor muscle, *2* neuromuscular junction, *3* synaptic vesicles, *4* degenerative corpuscle (×15,000)

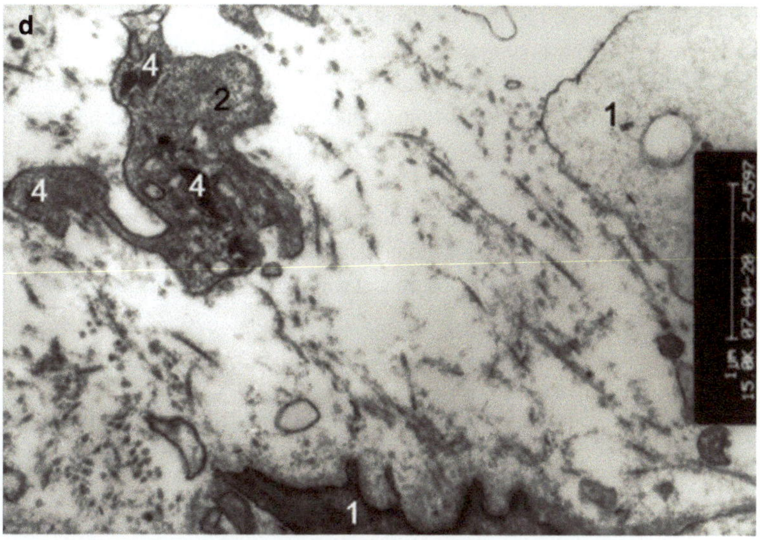

Fig. 4.2 (continued)

a high amount of degenerative corpuscles were observed (Fig. 4.3).

4.3.4 DM Fiber Properties

In group A, the muscular cells of the DM were bunched; the fibers showed similar courses and arrangement of the nucleus, with a small amount of collagen found distributed in the muscular bundles. By comparison, the muscle fibers in groups B–G showed gradual progression of fibrosis with time (Fig. 4.2). Obviously tiny muscular bundles, misalignment of smooth muscles, a large number of collagen fibers and infiltration of connective tissue among the muscular bundles were observed (Fig. 4.1). The percentage of connective tissue in the DM of groups A–G was 12.08 %±1.45 %, 18.47 %±1.32 %, 23.24 %±2.76 %, 26.29 %±1.88 %, 35.67 %±1.12 %, 47.22 %±2.38 % and 55.56 %±3.12 %, respectively. The percentage of connective tissue in groups E, F and G was significantly higher ($P<0.05$) than that in group A, but there were no significant differences between groups B, C and D, and group A.

4.3.5 Expression of α-SMA

Immunohistochemical staining demonstrated the α-SMA was distributed mainly in the DMs. A trans-section of the spinal cord showed lighter staining for α-SMA in the DM. The mean IOD for α-SMA in groups A–G was 0.96±0.03, 0.85±0.10, 0.74±0.06, 0.69±0.09, 0.44±0.08, 0.39±0.10, and 0.36±0.07, respectively. The mean IOD for α-SMA in groups E, F and G was significantly lower ($P<0.05$) than that in group A, but there was no significant difference between groups B, C and D and group A (Fig. 4.2).

4.4 Discussion

We were able to study in detail the structural changes that occur over time in the DM and NMJ after medullary cone injury in SD rats.

Motor end-plates and the NMJ of skeletal muscles have been widely studied; however, only a few studies have been conducted on smooth muscles, which have mainly been the focus of studies on gastrointestinal motility disorders and gastrointestinal tract tumors [13–15]. In particular, no study has focused on the DM and its NMJ. Previous studies have suggested that the NMJ of smooth muscles is different from that of skeletal muscles. Therefore, it is important to understand well the structure and functions of the NMJ of smooth muscles too.

The NMJ of smooth muscles without independent ending structures is distributed around the target muscle in the form of a thin neuroplexus; additionally, neurotransmitters are released and diffused from the nerve terminal, which might stimulate more adjacent muscular fibers [16–19]. These findings are consistent with those of our present study.

The DM is innervated by the parasympathetic nerve through a twice neuron exchange mode. So, in medullary cone injuries, the lowest level neurons localized in or near the bladder wall,

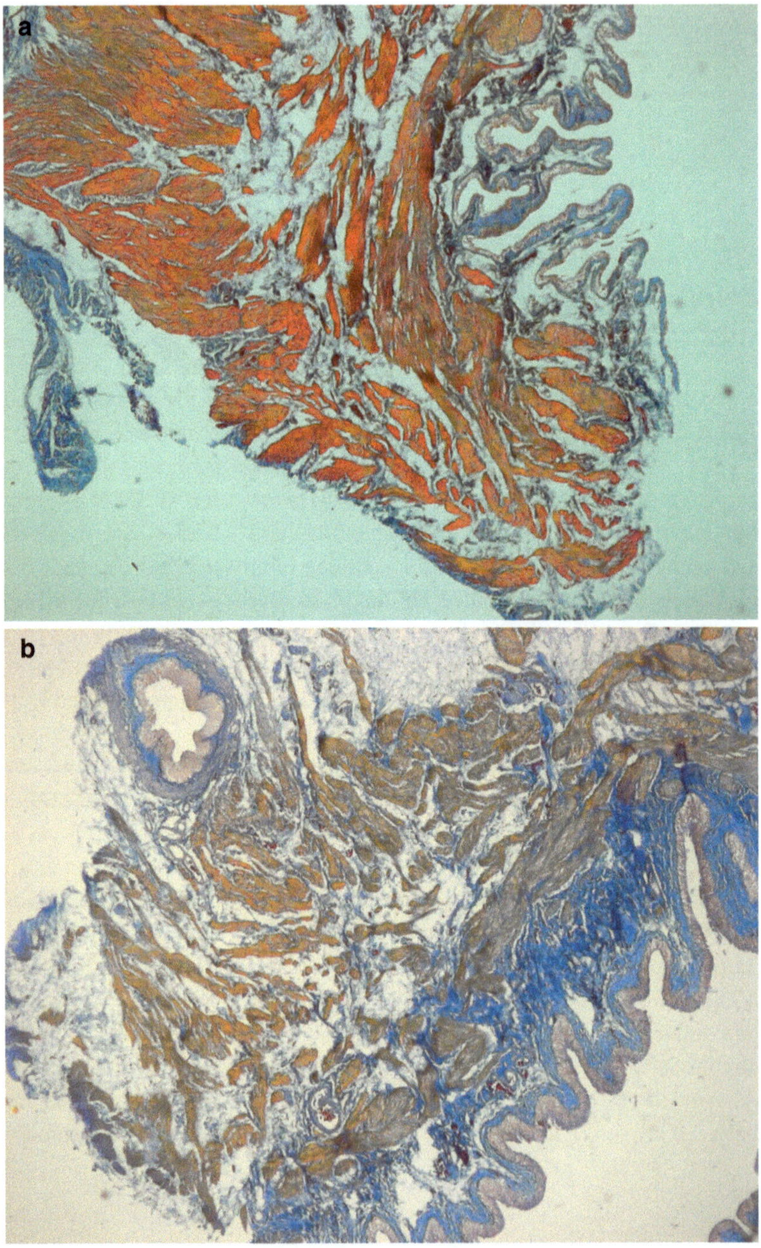

Fig. 4.3 Transverse sections of the detrusor muscle in rats of the normal control (**a**), and 6 weeks (**b**), 10 weeks (**c**) and 16 weeks (**d**) after medullary cone injury groups. Masson trichrome staining for analysis of fibrosis (×100)

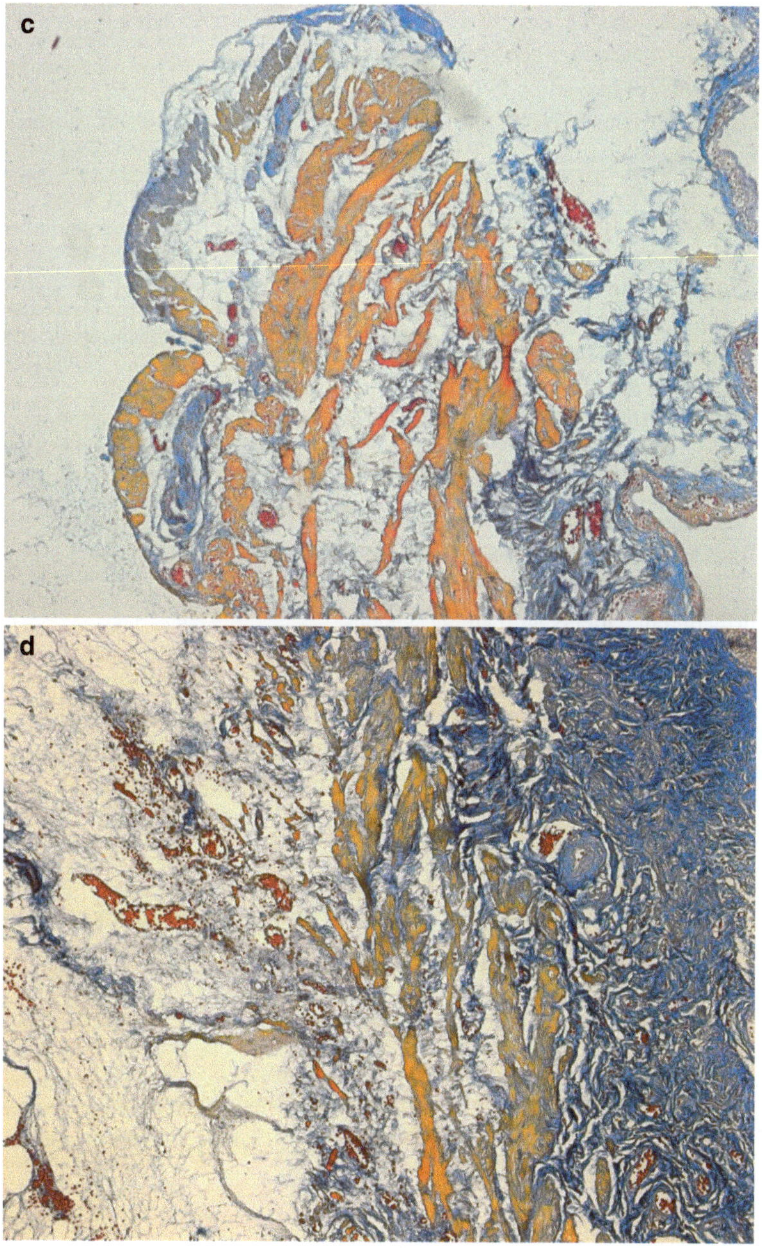

Fig. 4.3 (continued)

which may be partly associated with neurotrophy, are often unaffected. However, our present study has revealed that this neurotrophy of the lowest level neurons is limited.

We observed that obvious degeneration of the DM and its NMJ occurred in the 10th postoperative week; this means that the DM and its NMJ sooner or later undergo irreversible degeneration along with prolonging of the denervation. In the present study, we did not observe obvious degeneration of the NMJ: slight reduction in the amount of mitochondria and synaptic vesicles was observed within 8 weeks after the medullary cone injury in the rats, and from the 10th week, conspicuous degeneration of the NMJ appeared. In the 12th or 16th week after injury, obvious reduction or disappearance of synaptic vesicles and mitochondria even to degenerative corpuscles was noted.

In addition, we also measured the sectional areas of muscular fibers and the percentage of connective tissue in the DM. With time, the percentage of connective tissue also increased significantly. These results indicate that the gradual muscular atrophy and fibrosis of the DM were consistent with degeneration of the NMJ. Finally, immunohistochemical staining for α-SMA suggested that the contractibility of smooth muscles declined with the progression of denervation.

In conclusion, our data show that conspicuous changes of the DM and its NMJ can be observed from the 10th week after medullary cone injury in rats. Therefore, nerve repair should be performed before treating the injury. In addition, due to species differences between humans and rats, further studies should be performed to find the optimal operative time to restore self-controlled urination in patients with atonic bladder.

References

1. Benevento BT, Sipski ML. Neurogenic bladder, neurogenic bowel, and sexual dysfunction in people with spinal cord injury. Phys Ther. 2002;82:601–12.
2. Dogan S, Safavi-Abbasi S, Theodore N, Chang SW, Horn EM, Mariwalla NR, et al. Thoracolumbar and sacral spinal injuries in children and adolescents: a review of 89 cases. J Neurosurg. 2007;106:426–33.
3. Heary RF, Salas S, Bono CM, Kumar S. Complication avoidance: thoracolumbar and lumbar burst fractures. Neurosurg Clin N Am. 2006;17:377–88.
4. Blok BF, Karsenty G, Corcos J. Urological surveillance and management of patients with neurogenic bladder: results of a survey among practicing urologists in Canada. Can J Urol. 2006;13:3239–43.
5. Samson G, Cardenas DD. Neurogenic bladder in spinal cord injury. Phys Med Rehabil Clin N Am. 2007;18:255–74.
6. Chuang DC, Chang PL, Cheng SY. Root reconstruction for bladder reinnervation: an experimental study in rats. Microsurgery. 1991;12:237–45.
7. Sundin T, Carlsson CA. Reconstruction of severed dorsal roots innervating the urinary bladder. An experimental study in cats. II. Regeneration studies. Scand J Urol Nephrol. 1972;6:185–96.
8. Xiao CG, Godec CJ. A possible new reflex pathway for micturition after spinal cord injury. Paraplegia. 1994;32:300–7.
9. Lin H, Hou C, Chen A, Xu Z. Innervation of reconstructed bladder above the level of spinal cord injury for inducing micturition by contractions of the abdomen – to – bladder reflex arc. Neurosurgery. 2010; 66:948–52.
10. Lin H, Hou CL, Zhong G, Xie Q, Wang S. Reconstruction of reflex pathways to the atonic bladder after conus medullaris injury: preliminary clinical results. Microsurgery. 2008;28(6):429–35.
11. Hiersemenzel LP, Curt A, Dietz V. From spinal shock to spasticity: neuronal adaptations to a spinal cord injury. Neurology. 2000;54:1574–82.
12. Storch JS. Lumbar burst fracture associated with bowel, bladder, and sexual dysfunction: case study. J Neurosci Nurs. 2005;37:68–71.
13. Dahlberg A, Perttilä I, Wuokko E, Ala-Opas M. Bladder management in persons with spinal cord lesion. Spinal Cord. 2004;42:694–8.
14. Hirsch NP. Neuromuscular junction in health and disease. Br J Anaesth. 2007;99:132–8.
15. Kobayashi H, Yamataka A, Lane GJ, Miyano T. Disseminated mixed intestinal dysmotility (DMID): a new intestinal ganglion cell disorder? Pediatr Surg Int. 2005;21:883–8.
16. Pirker ME, Rolle U, Shinkai T, Shinkai M, Puri P. Prenatal and postnatal neuromuscular development of the ureterovesical junction. J Urol. 2007;177:1546–51.
17. Burnstock G. Innervation of bladder and bowel. Ciba Found Symp. 1990;151:2–18.
18. Fry CH, Hussain M, McCarthy C, Ikeda Y, Sui GP, Wu C. Recent advances in detrusor muscle function. Scand J Urol Nephrol Suppl. 2004;215:20–5.
19. Hashitani H, Bramich NJ, Hirst GD. Mechanisms of excitatory neuromuscular transmission in the guinea-pig urinary bladder. J Physiol. 2000;524:565–79.

5 Microstructural Study of the Detrusor Muscle After Repair of Atonic Bladders Caused by Injury to the Medullary Cone in Rats

Yuelei Zhang, Jun Sheng, and Chunlin Hou

5.1 Introduction

Thoracolumbar fractures often cause injury to the medullary cone, resulting in the loss of sensory and motor functions of the bladder, which is termed "atonic bladder". Severe urinary retention, refractory urinary-tract infection and the resultant chronic renal failure become the main cause of death in subjects with spinal cord injury (SCI) [1]. Hou et al. [2, 3] used the normal nerve roots above the level of the SCI as the donor nerves to reconstruct urinary function. They demonstrated reinnervation of the bladder through studies of electrophysiology, intravesical pressure and histology. Zheng et al. [4] studied the ultrastructural changes in the detrusor muscle and its neuromuscular junction (NMJ) after medullary-cone injury (MCI) in rats, and noted the features of degeneration and the optimal time of surgery.

We wanted to know if reconstruction of bladder function can prevent degeneration of the detrusor muscle. Hence, we studied microstructural changes in the detrusor muscle after repairing atonic bladders caused by MCI with normal nerve roots above the level of the SCI. Our primary purpose was to ascertain if reconstruction of urinary function with the normal lumbosacral nerve roots above the level of the SCI could constrain regression of the detrusor muscle.

Y. Zhang • J. Sheng • C. Hou, M.D (✉)
Department of Orthopedic Surgery, Changzheng Hospital, The Second Military Medical University, Fengyang Road 415, Shanghai 200003, People's Republic of China
e-mail: zhang_yue_lei@163.com; 449250924@qq.com; chunlin_hou@163.com

5.2 Material and Methods

5.2.1 Animal Preparation and Surgical Procedures

A total of 104 adult female Sprague–Dawley rats were randomized equally into 13 groups: normal; groups involving the repair of urinary function after MCI 1, 2, 3, 4, 5, and 6 min after surgery, which were assigned as groups T1–T6, respectively; and groups involving non-repair of urinary function after MCI 1, 2, 3, 4, 5, and 6 min after surgery, which were assigned as groups t1–t6, respectively.

Groups t1–t6 were created in a specific way. Briefly, rats were weighed and anesthetized (1 % sodium pentobarbital, 40 mg/kg), and placed prone on a table. After shaving the fur and sterilizing the incision site, laminectomy was undertaken at the L4/5 vertebral plane. The spinal cord was transected sharply at this plane with surgical scissors. The wound was sterilized and sutured after the procedure.

Groups T1–T6 were also created in a specific way. Briefly, rats were weighed and anesthetized and placed prone on a table. After shaving the fur and sterilizing the incision site, a longitudinal median incision was made from L4 to S4 to expose the cauda equina. The bilateral L5 ventral

C. Hou (ed.), *Functional Bladder Reconstruction Following Spinal Cord Injury via Neural Approaches*, DOI 10.1007/978-94-007-7766-8_5, © Springer Science+Business Media Dordrecht 2014

roots (VRs) were transected before they merged as mixed nerves. The S2 VRs were transected at the level of the L5 foramen intervertebrale. The bilateral L5 dorsal roots (DRs) were transected in the post-ganglionic area with the short mixed nerves. The S2 DRs were transected in the immediate post-ganglionic area. End-to-end anastomoses were carried out in L5 and S2 VRs. The DR stumps were coapted with a nerve graft (the S2 DRs in the dural sac with the exception of dorsal root ganglia). All nerves were repaired with one stitch using 12–0 monofilament nylon under a microscope at × 10 magnification. After the procedure, penicillin (400,000 U/day) was administered for 3 consecutive days to prevent infection. Rats in the injury groups and repair groups could not urinate by themselves. Urination was achieved by pressing upon the abdomen once a day. Rats that died were replaced with other rats to ensure that the sample size in each group remained constant.

5.2.2 Residual Urine Volume

At the designated time, the abdomens of rats in each group were squeezed to achieve urination at a certain time in the morning of 3 consecutive days, and the urine volume recorded. In addition, the urine volume of rats in the repair group and injury group were recorded 1–3 days after the procedure, and assigned as T0 and t0, respectively.

5.2.3 Histological Observation of the Detrusor Muscle

The detrusor muscle was fixed in 4 % paraformaldehyde solution for 48 h. It was then dehydrated and immersed in paraffin wax. The middle of the detrusor muscle was sliced into 5-μm sections using a LKB-I Ultramicrotome (LKB, Bromma, Sweden). After dyeing with hematoxylin and eosin (H&E), morphological changes were observed under a light microscope. Three fields of each section were analyzed using the Qwin300 Digital Imaging Workstation to calculate the mean cross-sectional area of longitudinal smooth-muscle fibers. The mean value of the three fields was recorded.

5.2.4 Electron Microscopic Observation of the Ultrastructure of the Detrusor Muscle and NMJ

The detrusor muscles of rats of the normal, T1, T3, T6, t1, t3 and t6 groups were pruned to tissue blocks of size <1 mm^3. They were then double-fixed in 2.5 % glutaraldehyde and osmic acid solution. They were then dehydrated with ethanol and acetone, embedded with Epon 812 resin (Electron Microscopy Sciences, Hatfield, PA, USA) and sliced using a LKB-I Ultramicrotome (LKB) into 50-nm sections. They were double-dyed with lead and uranium and observed using a Philips CM120 Transmission Electron Microscope (Philips, Amsterdam, the Netherlands) for ultrastructural observation of the detrusor muscle and NMJ (mainly changes in synaptic vesicles and organelles).

5.2.5 Analyses of Fibrotic Components in the Detrusor Muscle

Sliced 5-μm paraffin sections of the detrusor muscle were dyed with Masson's trichrome stain. Using this stain, collagen fibers are blue, muscle fibers and the cytoplasm are red, and the nucleus blue and brown. Sections were analyzed using the Qwin300 Digital Imaging Workstation to calculate the percentage of connective tissue in detrusor muscle using a magnification of × 20.

5.2.6 Statistical Analyses

Data are the mean and standard deviation. Two-way ANOVA was used to determine the difference between the groups of MCI and bladder-function reconstruction. One-way ANOVA was used to determine intragroup differences. The least-square difference method was employed for multiple comparisons. $P<0.05$ was considered significant. Data were analyzed using SPSS ver13.0 (SPSS, Chicago, IL, USA).

5.3 Results

5.3.1 Residual Urine Volume

A significant difference was observed between the groups of MCI and bladder-function reconstruction. In t1–t6, the residual urine volume increased with time and there were significant differences among these groups ($P<0.05$). An increase in the residual urine volume was observed in T1–T3 but decreased in T4–T6 (Table 5.1).

Table 5.1 Residual urine volume after surgery ($n=8, \bar{x}\pm s$)

Reconstruction group	Residual volume (mL)	Injury group	Residual volume (mL)
Normal	3.31±0.18	Normal	3.28±0.20
T1	5.09±0.19	t1	5.06±0.16
T2	7.54±0.18	t2	7.78±0.12
T3	9.85±0.29	t3	12.03±0.18
T4	6.35±0.23	t4	15.11±0.30
T5	4.11±0.16	t5	18.22±0.37
T6	2.31±0.22	t6	20.15±0.53

5.3.2 Histological Observation of the Detrusor Muscle

In the normal groups, the smooth muscle cells (SMCs) were in a long "shuttle" shape and funicular in arrangement. The nucleus was arranged in an orderly manner and parallel to muscle cells. Infiltration into connective tissue among the muscle bundles was not observed.

In t1–t6 and T1–T3, muscle cells were arranged in an irregular manner and the nucleus was deformed. The detrusor muscle was smaller, and increased infiltration into connective tissue was observed with time.

In T4–T6, deformed muscle cells were seen, but the area of detrusor muscle increased and increased infiltration into connective tissue was not observed (Fig. 5.1).

Study of the mean-cross sectional area of the detrusor muscle suggested that it decreased significantly in t1–t6 with time and there were significant differences among the groups of MCI. However, in the reconstruction groups, the mean-cross sectional area of the detrusor muscle decreased in T1–T3 but increased in T4–T6 (Table 5.2).

5.3.3 Ultrastructure of the Detrusor Muscle and NMJ

In the normal groups, SMCs were in a long shuttle shape and aligned in an orderly manner. The intercellular space was small and uniform. Mitochondria and endoplasmic reticula were not swollen. Dense bodies and parallel-arranged myofilaments could be seen in muscle cells. Many pinocytotic vesicles were observed under the muscle membrane. A normal NMJ was seen between the cells of the detrusor muscle, with abundant synaptic vesicles, mitochondria and endoplasmic reticula.

In the reconstruction and injury groups, atrophy of muscle cells, disorganized fibers, and decreased/absent connections between cells were observed. An increased distance between cells and an appreciable number of collagenous fibers between myocytes were noted. Intracellular organelles were swollen and the number of myofilaments reduced. Dense bodies and pinocytotic vesicles in myocytes were observed. A reduction in the number of synaptic vesicles in the NMJ of T1 and T3 was noted, but more synaptic vesicles were observed in the NMJ of T6. In t1, the number of synaptic vesicles in the NMJ was less than that in the NMJ of the normal group. In t3, the structure of the NMJ degenerated with a deficiency of synaptic vesicles. In t6, the NMJ was rarely seen between detrusor-muscle cells (Fig. 5.2).

5.3.4 Analyses of Connective Tissue

In the normal groups, detrusor-muscle cells were aligned and little connective tissue was observed among the muscle bundles. In t1–t6, fibrosis developed with time along with tiny muscle bundles. Infiltration of connective tissue was seen in T1–T6, but was relatively stable in T3–T6 (Fig. 5.3) (Table 5.3).

5.4 Discussion

Three nerves innervate the bladder. The pelvic nerve originates from S2 to S4, consists of parasympathetic nerve fibers, and governs contraction of the detrusor muscle. The hypogastric nerve originates at T11–L2, consists of sympathetic nerve fibers, and inhibits voiding of the detrusor muscle. The pudendal nerve contains

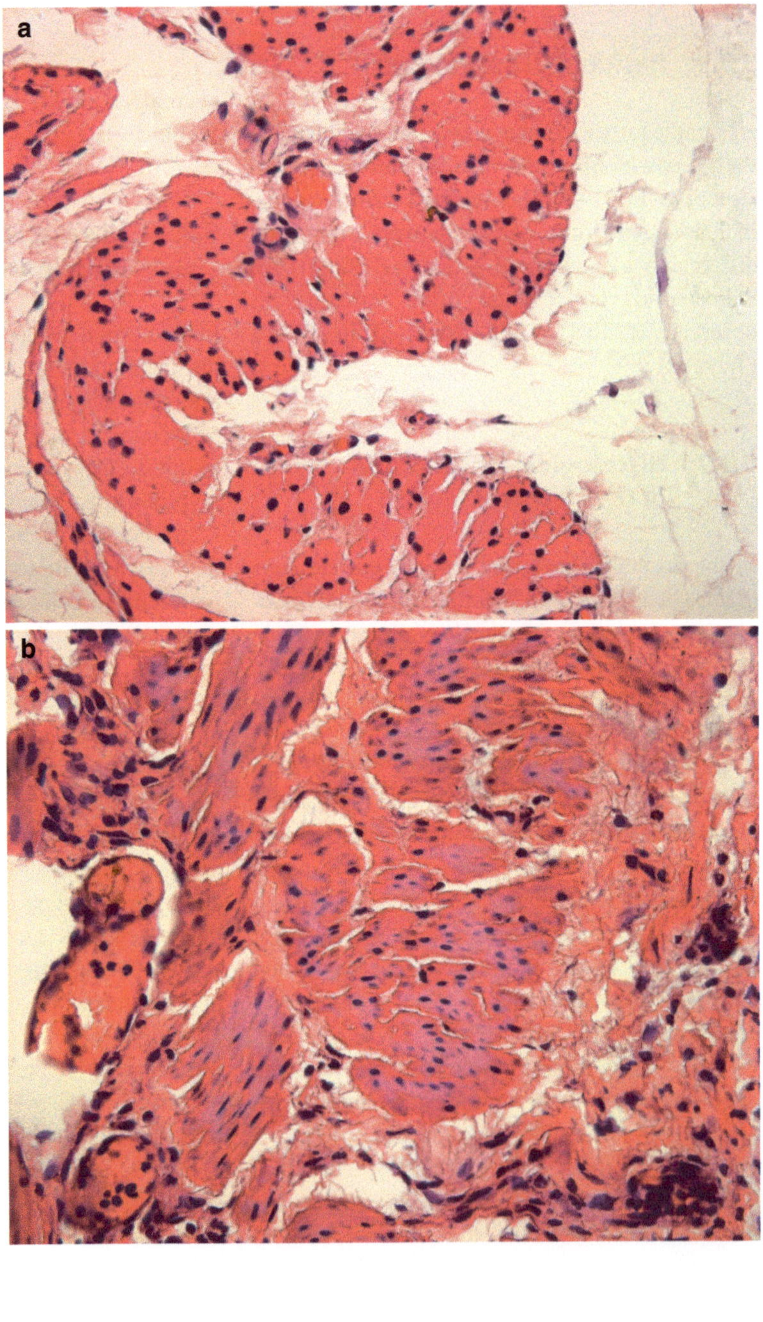

Fig. 5.1 Transverse sections of the detrusor muscle in Sprague–Dawley rats of normal (**a**) medullary-cone injury (**b**–**g**) and reconstruction (**h**–**m**) groups stained by hematoxylin and eosin for analyses of the cross-sectional areas of muscle fibers (×400 magnification)

Fig. 5.1 (continued)

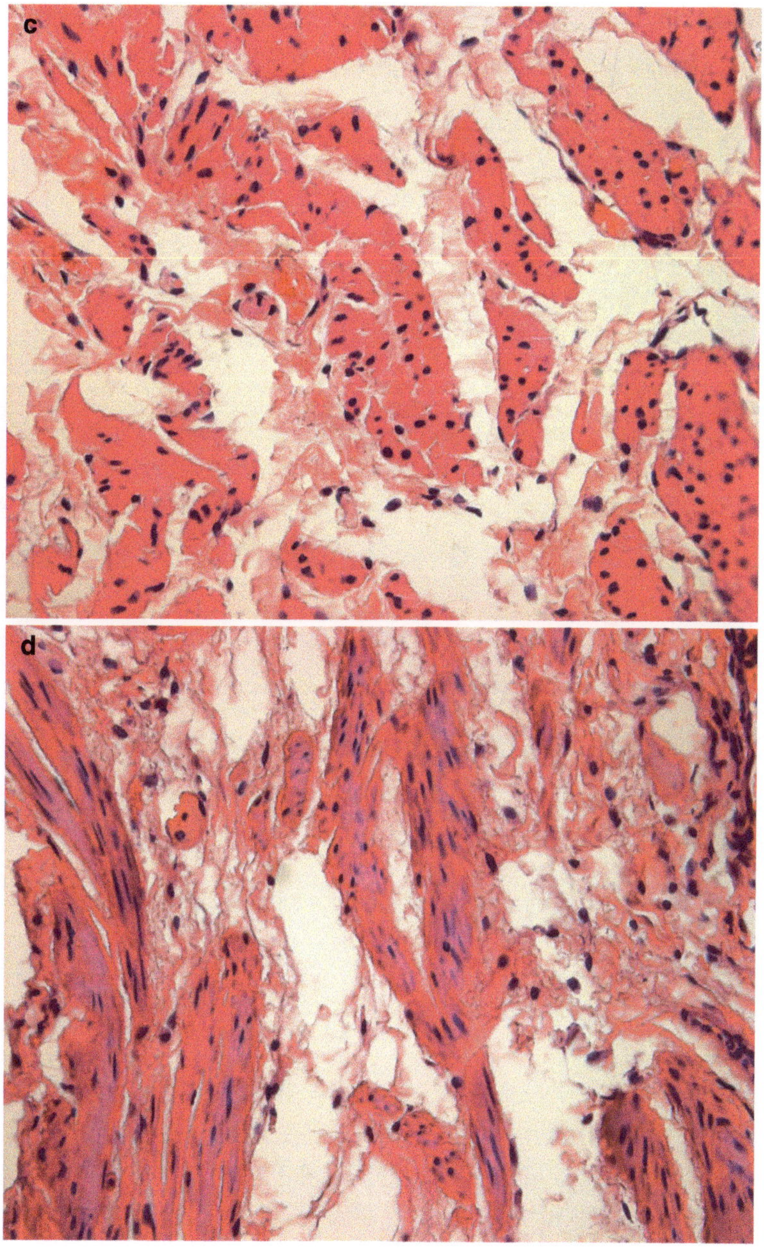

Fig. 5.1 (continued)

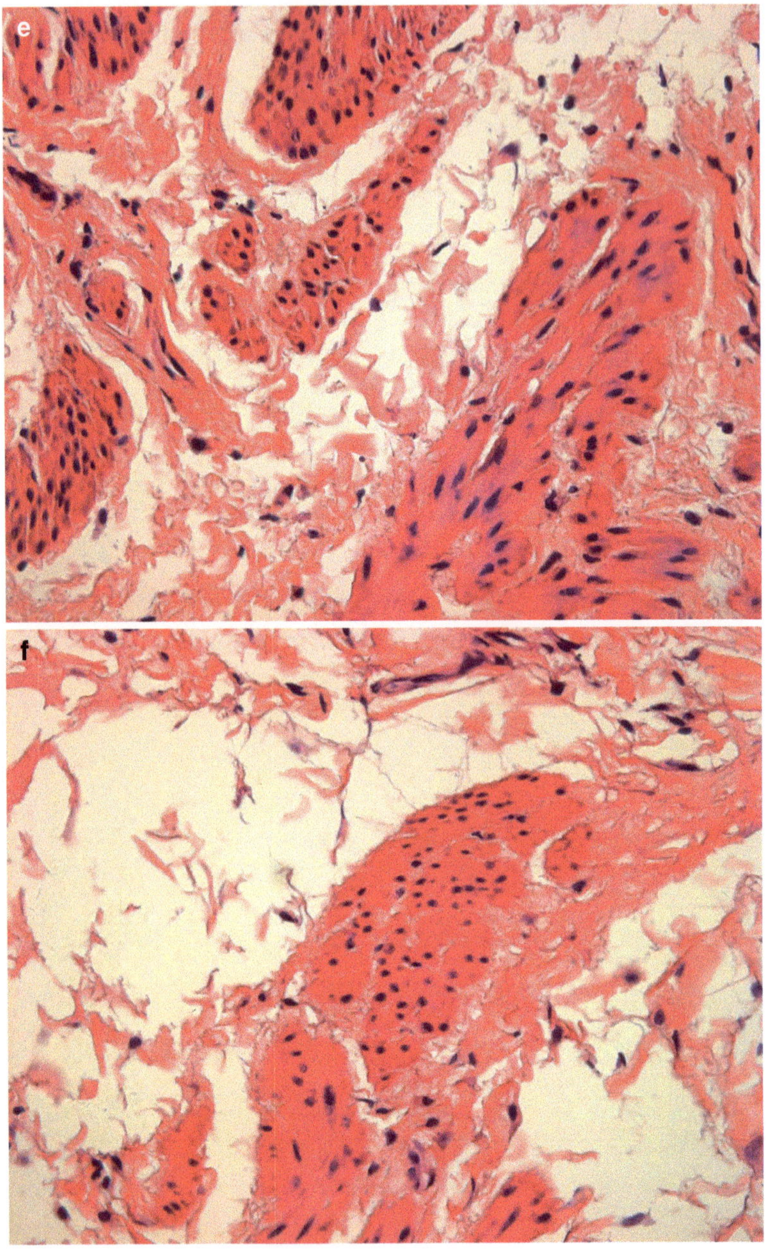

Fig. 5.1 (continued)

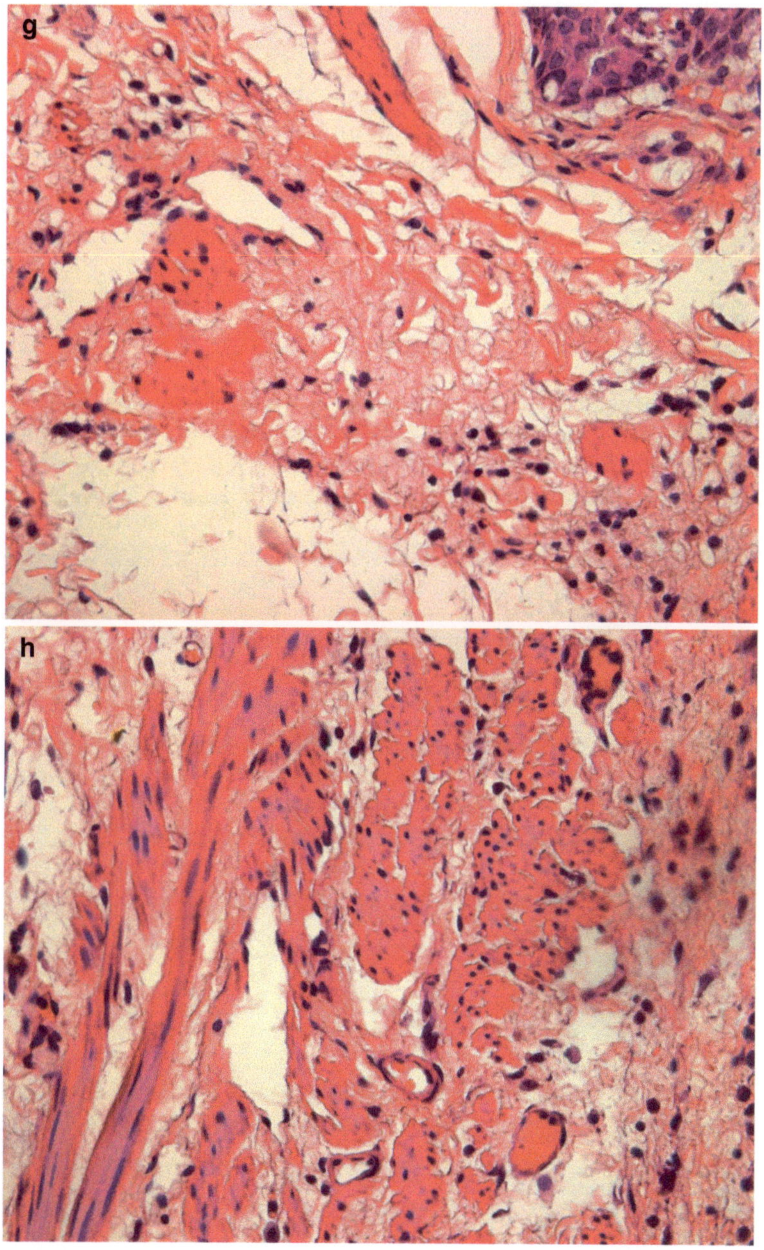

Fig. 5.1 (continued)

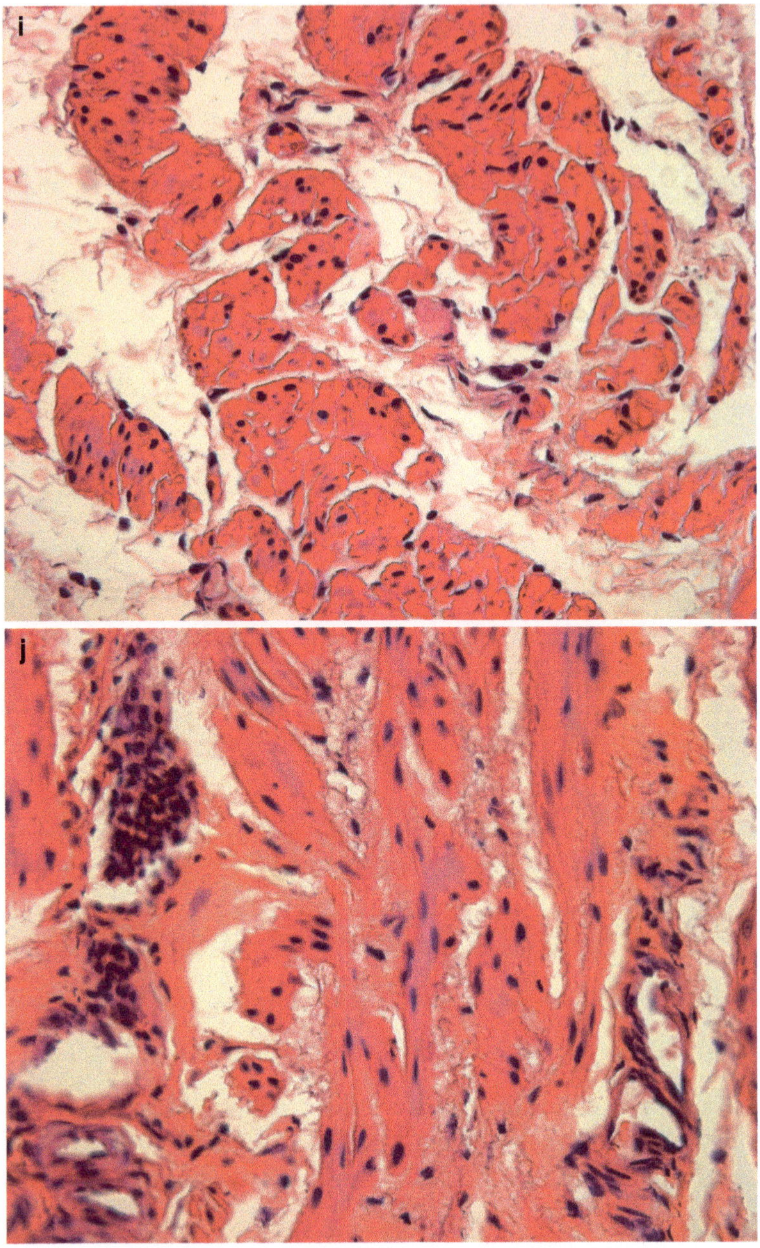

Fig. 5.1 (continued)

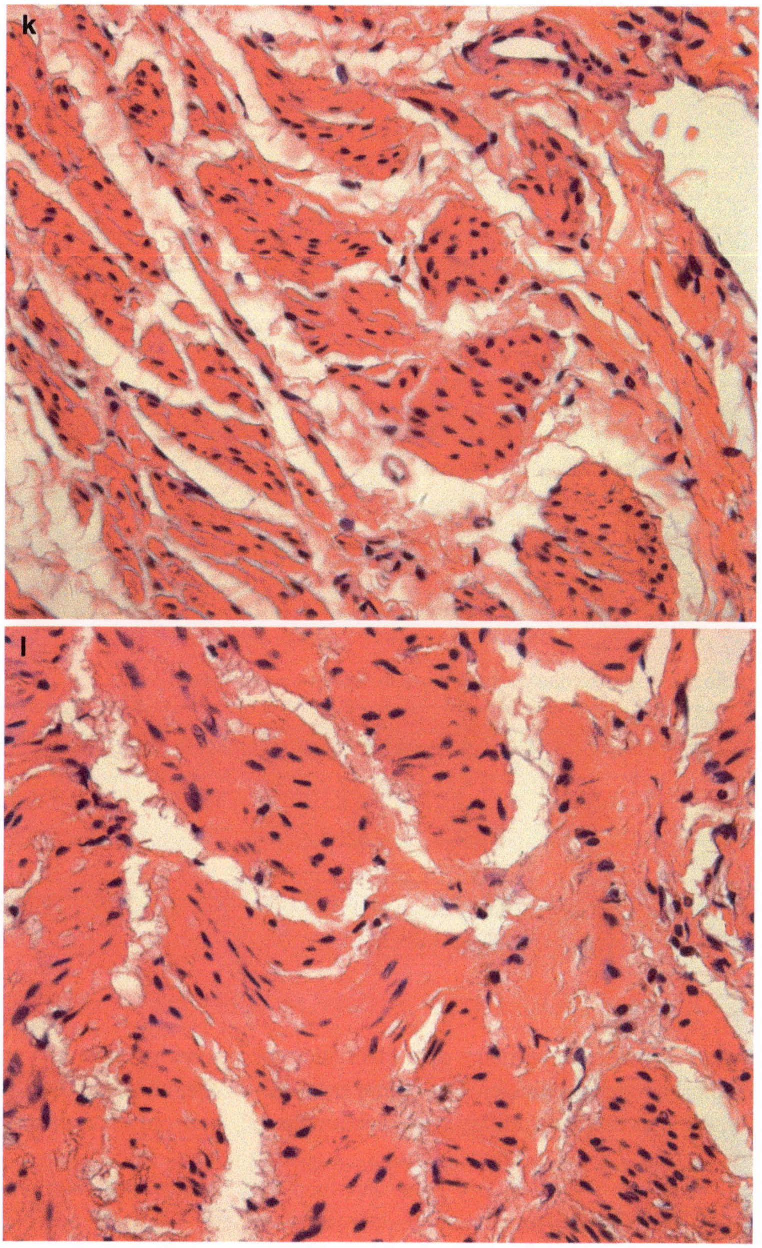

Fig. 5.1 (continued)

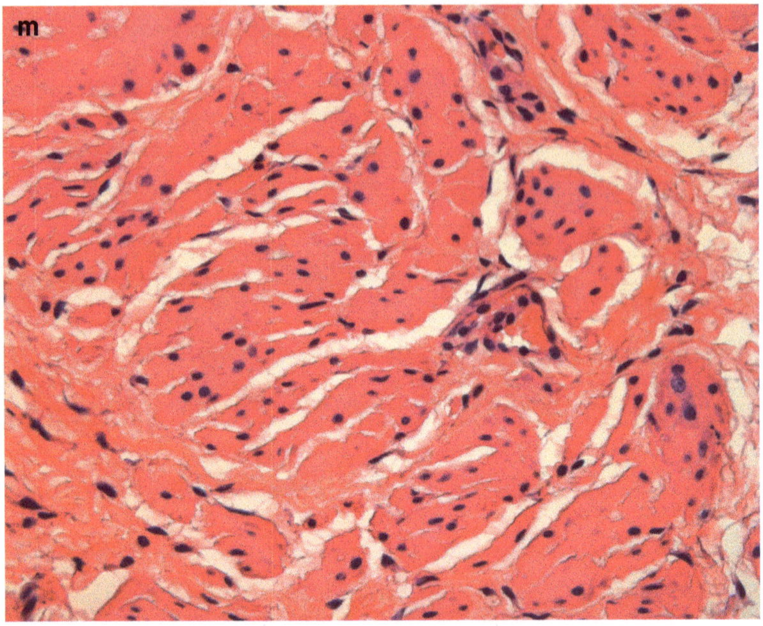

Table 5.2 Cross-sectional area of the detrusor muscle (n = 8, $\bar{x} \pm s$)

Reconstruction group	Mean cross-sectional area (μm^2)	Injury group	Mean cross-sectional area (μm^2)
Normal	6,163.13 ± 583.50	Normal	6,163.13 ± 583.50
T1	4,964.04 ± 242.09	t1	4,994.7 ± 183.11
T2	3,800.47 ± 283.54	t2	3,748.8 ± 208.91
T3	3,341.43 ± 285.66	t3	2,964.8 ± 182.89
T4	4,432.19 ± 401.26	t4	2,047.22 ± 219.30
T5	4,680.22 ± 248.00	t5	1,483.98 ± 133.93
T6	5,246.43 ± 343.71	t6	962.01 ± 110.09

mainly somatic nerve fibers and innervates the external urethral sphincter. The sacral cord S2–S4, one part of the medullary cone, is a low-level micturition center. Thoracolumbar fractures often damage the medullary cone and then the reflex arc, leading to an acontractile detrusor muscle, strained sphincters and, eventually, an atonic bladder [5].

In health, the detrusor muscle comprises spindle-shaped SMCs with a single nucleus. The muscle bundles are surrounded by thick interstitial macrosepta of collagen, elastin and, occasionally, fibroblasts. Under electron microscopy, numerous mitochondria and, occasionally, endoplasmic reticula and the Golgi apparatus are observed. In the sarcoplasm, clusters of ribosomes and dense bodies are seen, along with rows of caveolae throughout the sarcolemma. The axon bundles are sparse in the interstitium and only a portion of myocytes can be stimulated directly by the NMJ; most receive the stimulus of contraction via mechanical coupling or electrical coupling in intercellular junctions [6]. Contraction of the detrusor muscle is based mainly upon three components: SMCs, the interstitium and intramural nerves. Myocyte morphology directly determines contraction of the detrusor muscle. Intramural nerves are mainly autonomic nerves, affecting contraction and relaxation through respective neurotransmitters [7]. However, the ends of these nerves surround the target muscle in the form of a neuroplexus without independent-ending structures, and the neurotransmitters released from the nerve terminals stimulate more adjacent muscular fibers in a diffusive way [8].

Ultrastructural changes in the detrusor muscle affect bladder function. Hindley et al. [9] examined 21 bladder samples of patients with severe hypocontractile detrusor muscles. They found: widespread disrupted muscle cells; sarcoplasmic vacuolation; disruption of cell organelles (and, ultimately cell lysis); and excessive deposits of collagen and elastic fibers between widely separated muscle cells and in the interstitium. MCI can cause an atonic bladder, but there are similar changes in the detrusor muscle apart from denervation: irregularly arranged muscle cells, fewer intermediate junctions with collagen disposition, decreased caveolae throughout the sarcolemma, swollen mitochondria in myocytes and chaotically arranged myofilaments [10]. These changes often bring poor contraction of muscle cells and poor coupling among myocytes. Zheng et al. [4] created a model of MCI through transection of the L4/5 vertebral plane. They found that conspicuous changes of the detrusor muscle and its NMJ occurred from the tenth week after MCI in rats and that this degeneration was aggravated with time. Severe degeneration can seriously

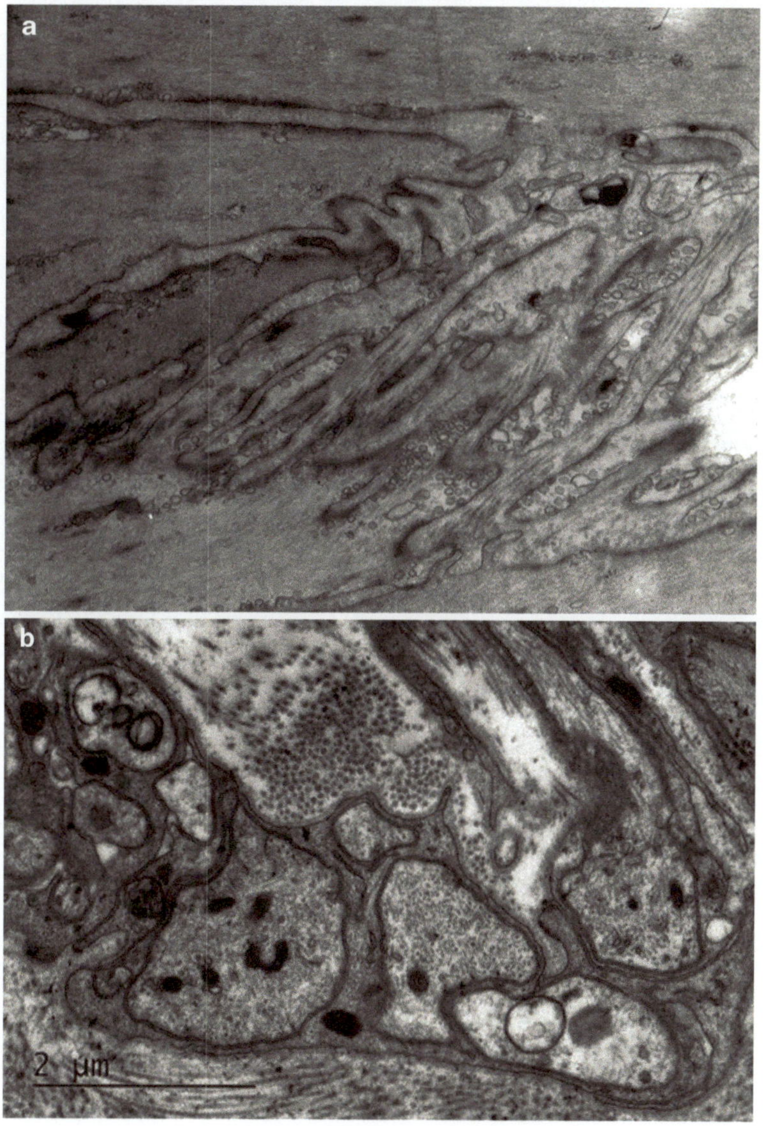

Fig. 5.2 Transmission electronic microscope observation of the detrusor muscle and neuromuscular junction (*NMJ*). A reduction in the number of synaptic vesicles in the NMJ of rats of T1 (**a**) and T3 (**b**) was noted, but more synaptic vesicles were observed in the NMJ of rats of T6 (**c**). In t1 (**d**), the number of synaptic vesicles in the NMJ was less than that in the normal group (**e**). In t3 (**f**), the structure of the NMJ degenerated with a deficiency of synaptic vesicles. In t6 (**g**), the NMJ was rarely seen between detrusor-muscle cells

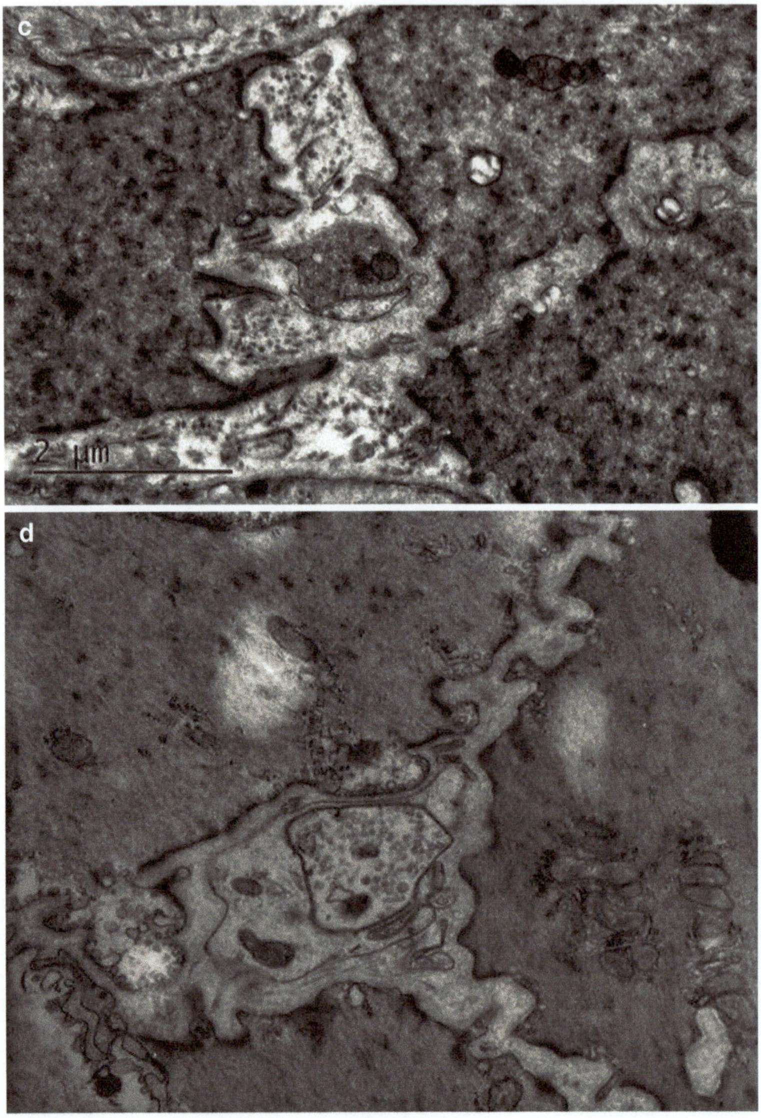

Fig. 5.2 (continued)

Fig. 5.2 (continued)

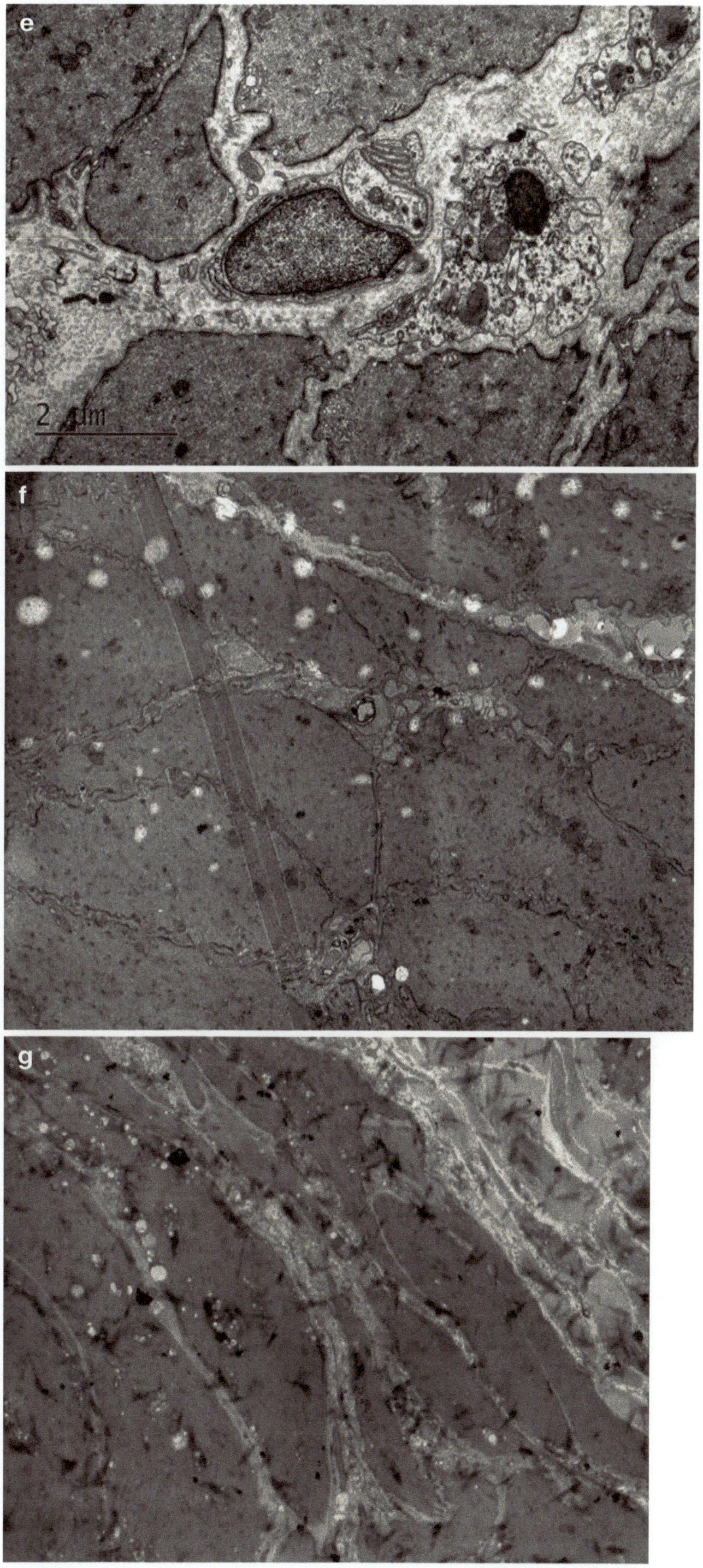

affect the recovery of an atonic bladder, so we wondered if early treatment would be helpful.

There are several treatments for an atonic bladder caused by MCI: clean intermittent catheterization (CIC) [11], abdominal compression [12], and drug therapy [13]. CIC is the "gold standard", but the procedure is not easy to carry out, and can lead to urethral injury and urinary tract infections. Therefore, "auto-urination" has become a research focus. Hou et al. [14] reconstructed the atonic bladder using the abdominal reflex remaining above the level of the SCI. They anastomosed the T11 VR to the S2 VR unilaterally through a nerve graft. Postoperatively, these patients voided urine by scratching or gently squeezing the T11 dermatome. Between 2007 and 2009, they used the normal S1 VR above the level of the SCI to innervate the atonic bladder

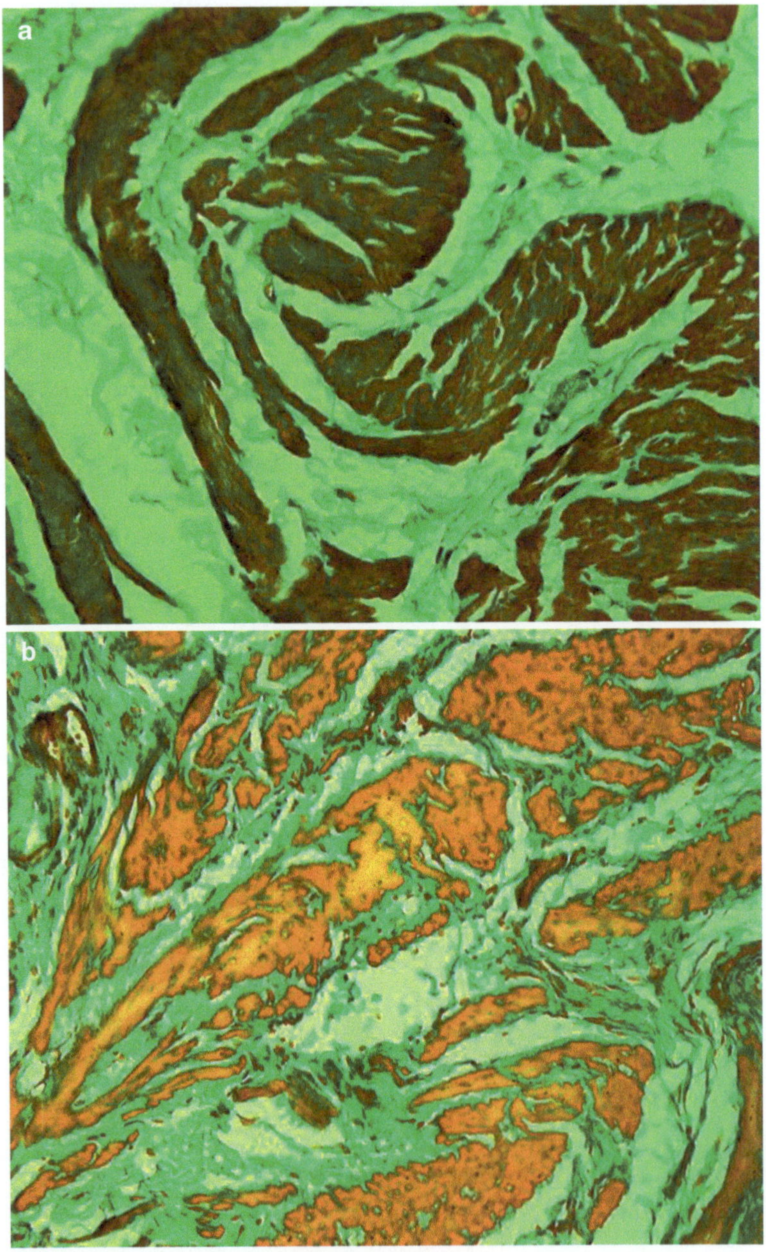

Fig. 5.3 Transverse sections of the detrusor muscle in Sprague–Dawley rats of normal (**a**) medullary-cone injury (**b–g**) and reconstruction (**h–m**) groups stained with Masson's trichrome for analyses of the connective tissue (×200 magnification)

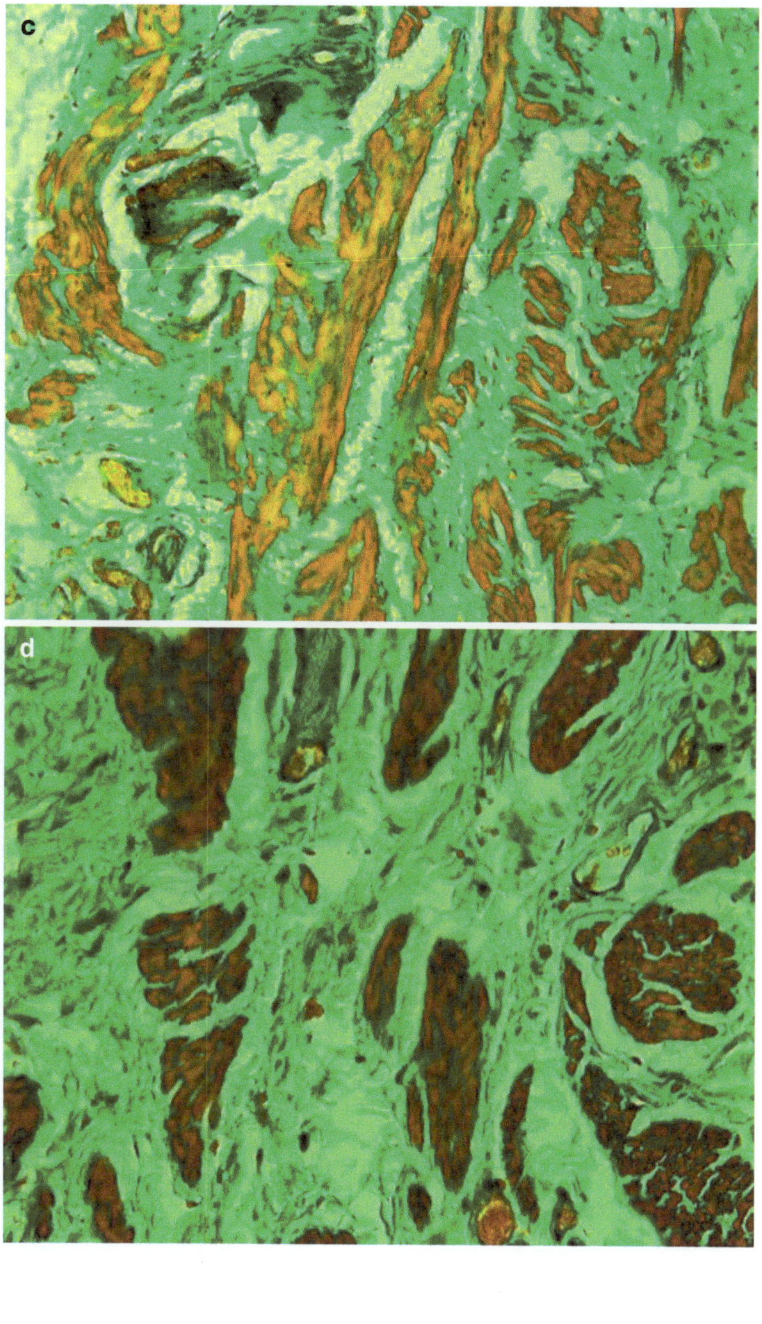

Fig. 5.3 (continued)

Fig. 5.3 (continued)

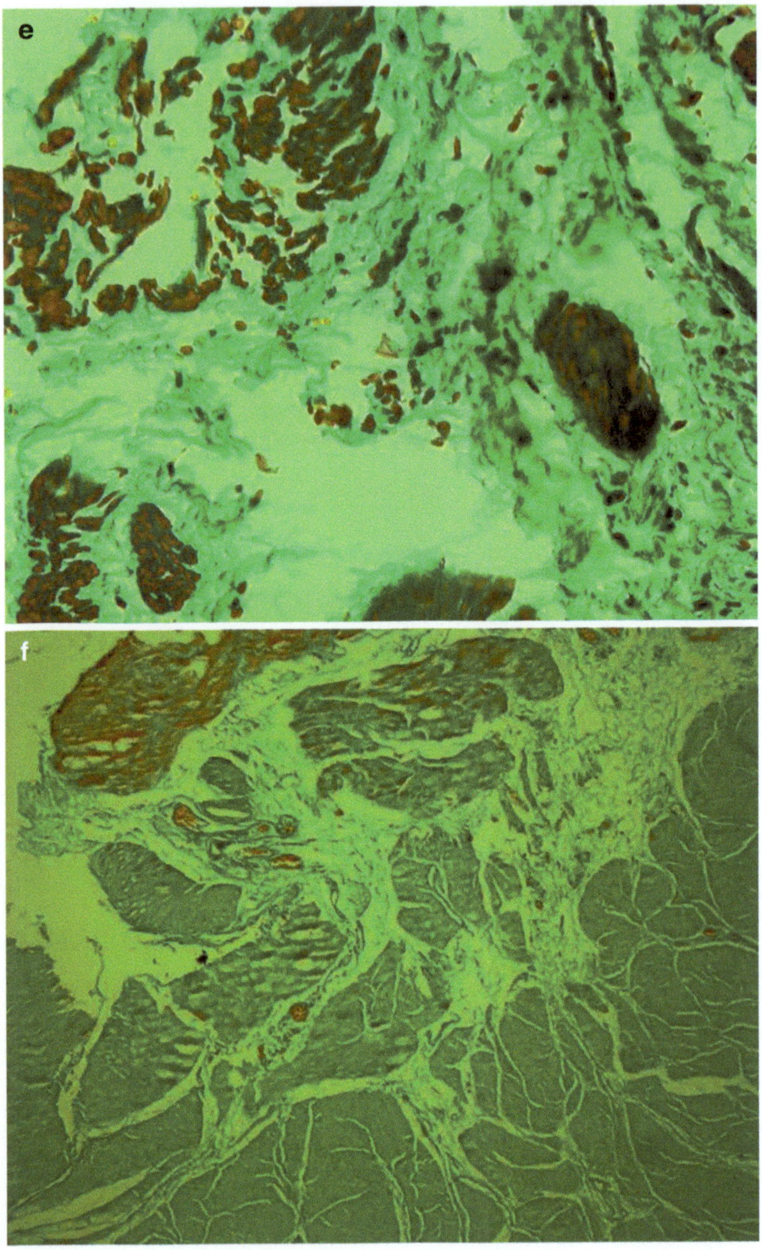

Fig. 5.3 (continued)

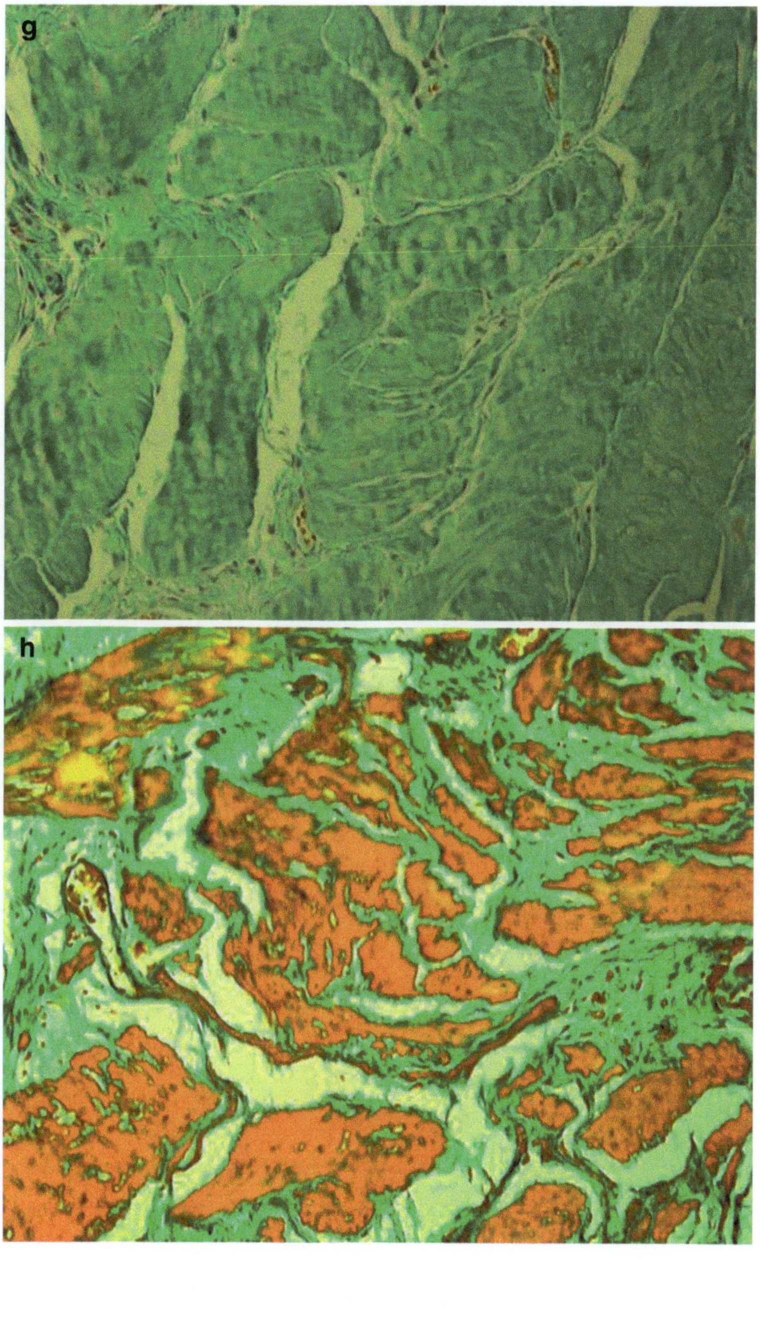

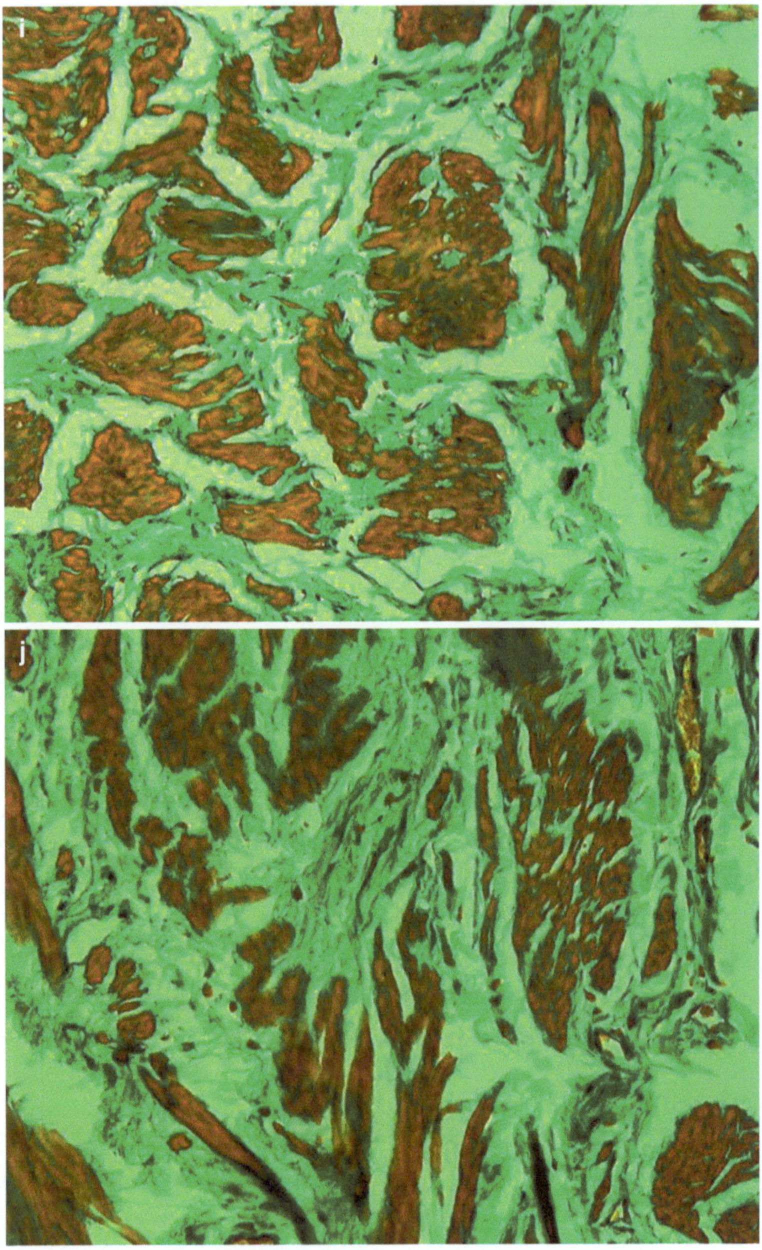

Fig. 5.3 (continued)

Fig. 5.3 (continued)

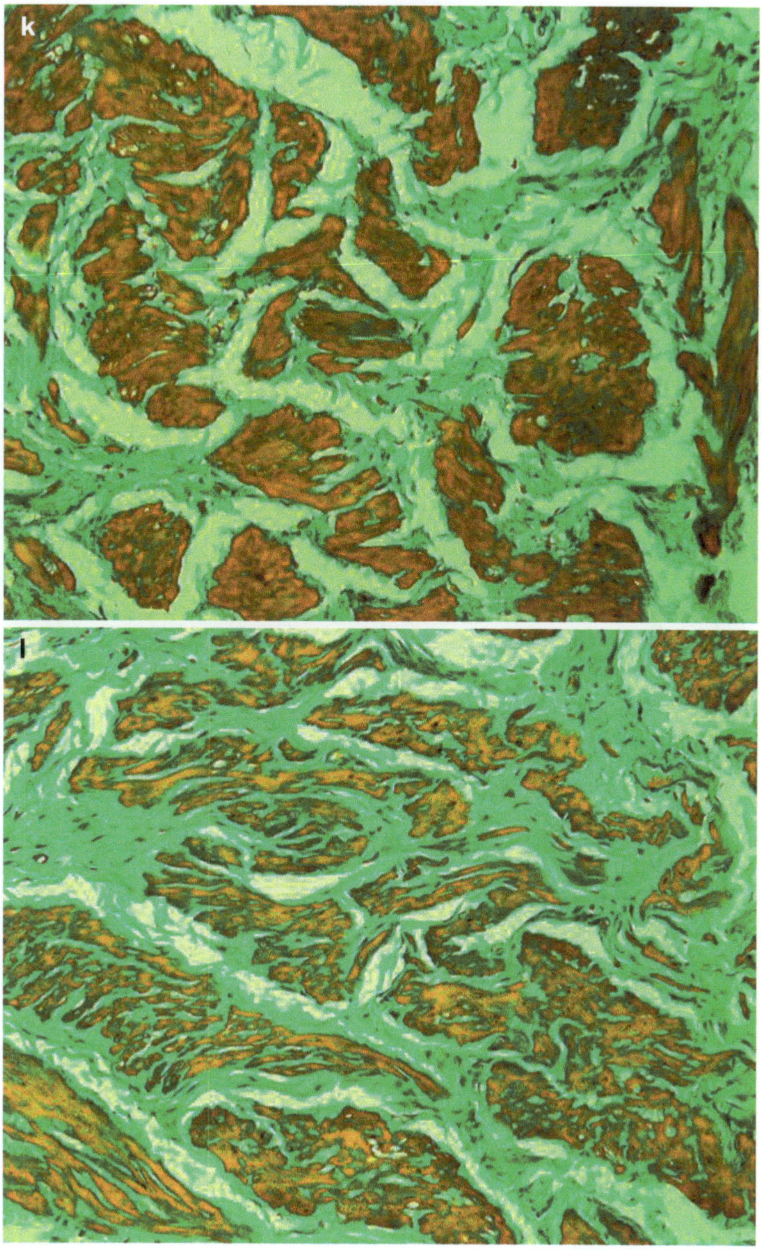

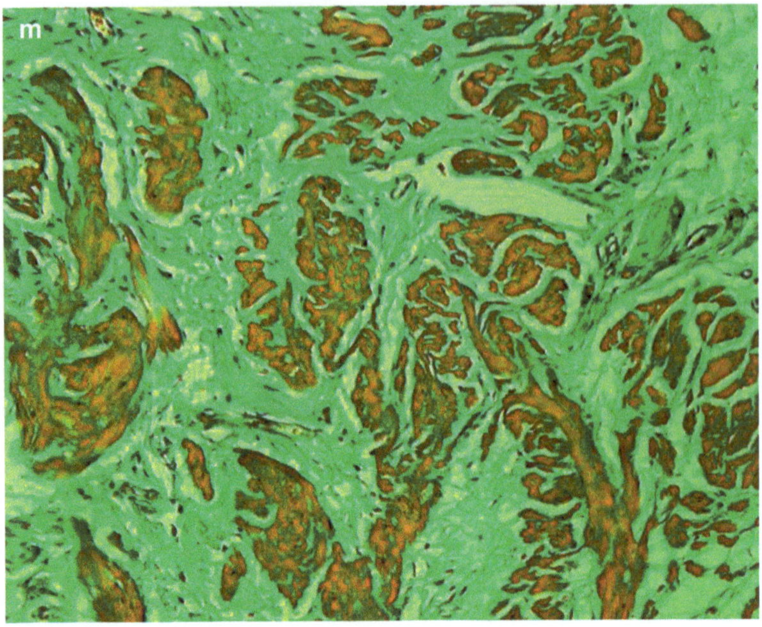

Fig. 5.3 (continued)

Table 5.3 Proportion of connective tissue in each group

Reconstruction group	Connective tissue (%)	Injury group	Connective tissue (%)
Normal (0 m)	12.83 ± 1.12	Normal (0 m)	12.83 ± 1.12
T1	21.64 ± 1.83	t1	20.40 ± 1.53
T2	26.16 ± 2.20	t2	26.78 ± 1.25
T3	29.01 ± 1.37	t3	47.34 ± 1.34
T4	29.40 ± 1.74	t4	72.93 ± 2.09
T5	30.29 ± 1.74	t5	84.62 ± 1.62
T6	32.06 ± 3.29	t6	93.54 ± 2.23

with normal lower extremities through neuroanastomosis with S2/3 VRs of the same side. Seven out of nine patients regained satisfactory bladder control within 8–12 months [13]. Lin et al. [15] constructed a new "sensor–sensor, motor–motor" reflex pathway by transfer of the L5 VR to the S2 VR root together with post-ganglionic spinal nerve transfer of the L5 DR to the S2 DR with a nerve graft in a SD mouse model (i.e., spontaneous mutation of the house mouse). They demonstrated the reconstruction of the sense and motor function of the bladder through studies of the electrophysiology, intravesical pressure and histology of the anastomoses. Reconstruction of urinary function can realize autonomous voiding in patients and has few complications, so it has good prospects.

In the present study, we constructed the sensor–sensor, motor–motor reflex pathway in Sprague–Dawley rats with MCI. The residual urine volume increased in the first 3 months, then decreased, after surgery. These findings suggested that the voiding function of rats had improved. We observed an increase in the number of synaptic vesicles in the NMJ of rats of the T6 compared with those of T3 by electron microscopy. This finding was a symptom of nerve regeneration, suggesting innervation of the atonic bladder. The increase in the mean cross-sectional area of the detrusor muscle 3 months after surgery reflected the improvement in the detrusor muscle with innervation. We showed that the proportion of connective was not significantly different in T3, T4 and T5, and that fibrosis of the detrusor muscle ceased upon reinnervation.

These data suggest that reconstruction of urinary function with normal lumbosacral nerve roots above the level of the SCI can restore voiding function and constrain the regression of the detrusor muscle.

References

1. Heary RF, Salas S, Bono CM, et al. Complication avoidance: thoracolumber and lumbar burst fractures. Neurosurg Clin N Am. 2006;17:377–88.
2. Zhong G, Hou C, Xie Q, et al. Experimental study on the artificial bladder reflex arc established in therapy of flaccid bladder after spinal cord injury. Chinese J Rep Reconstr Surg. 2006;20:812–5.
3. Lin H, Hou C. Transfer of normal S1 nerve root to reinnervate atonic bladder due to conus medullaris injury. Muscle Nerve. 2013;47:241–5.
4. Zheng XY, Hou CL. Optimal timing of operation for repairing atonic bladder after medullary cone injury: an experimental study in rats. Spinal Cord. 2008;46: 574–81.
5. Swartz D. The neurogenic bladder in spinal cord injury. Spinal Cord Inj. 1946;64:333–9.
6. Elbadawi A. Functional anatomy of the organs of micturition. Urol Clin North Am. 1996;23:177–210.
7. Elbadawi A. Pathology and pathophysiology of detrusor in incontience. Urol Clin North Am. 1995;22:499–512.
8. Hou C, Zheng X. Experimental study on morphological changes of detrusor muscle and its neuromuscular junction after medullary cone injury in rats. Spinal Cord. 2007;21:473–7.
9. Hindley RG, Brierly RD. A qualitative ultrastructural study of the hypocontractile detrusor. J Urol. 2002;168:126–31.
10. Zhao JM, Zhang YH. Ultrutructural observations on detrusor musculature in areflexia neurogenic bladders. China J Urol. 2004;25:682–4.
11. Emmanuel CK, Pierre D. Intermittent catheterization with hydrophilic catheters as a treatment of chronic neurogenic urinary retention. NeurourolUrodyn. 2011;30:21–31.
12. Chang SM, Hou CI, Dong DQ, et al. Spinal cord injury patients from the 1976 Tangshan earthquake, and managed for over 20 years using the Crede m aneuver. Spinal Cord. 2000;38:552–4.
13. Wein AJ. Pharmacologic therapy for incontinence. Urol. 1990;36:36–43.
14. Lin H, Hou CL, Zhong G, et al. Reconstruction of reflex pathways to the atonic bladder after conus medullaris injury: preliminary clinical result. Microsurgery. 2008;28:429–35.
15. Lin H, Hou C, Zhen X. Bypassing spinal cord injury: surgical reconstruction of afferent and efferent pathways to the urinary bladder after conus medullaris injury in a rat model. J Reconstr Microsurg. 2008;24:575–81.

Selective Sacral Rhizotomy: Introducing a Simple Intraoperative Manometric Method

6

Chunlin Hou and Mingxuan Liu

6.1 Background

Spinal cord injury not only results in somatic nerve paralysis but also makes the visceral organs lose higher central control and dominance, causing neural dysfunction of the sex organs. Injury above the cone accounted for 75 % of all kinds of spinal cord injury. After recovery, most patients will develop a spastic bladder with high tension and reflectance. Urinary incontinence and failure of urinary storage may exist at the same time. Not only is there a barrier in the functions of both urine storage and micturition, but also the presence of high pressure bladder causes ureteral reflux and consequent damage to the kidney, which is the primary cause of death of paraplegic patients. Therefore, improvement of the bladder functions of urine storage and micturition is the basic premise behind attempts to prolong patients' life and improve its quality. Based on our success in experimental research on selective sacral rhizotomy in the treatment of spastic bladder after spinal cord injury, we started to carry out clinical operations on some cases in 1998 [1, 2]. So far we have completed 32 cases, among which the longest follow-up is 33 months, all exhibiting a prominent treatment effect.

C. Hou, M.D. (✉) • M. Liu
Department of Orthopedic Surgery, Changzheng Hospital, The Second Military Medical University, Fengyang Road 415, Shanghai 200003, People's Republic of China
e-mail: chunlin_hou@163.com; 1988001@sina.com

6.2 Animal Experiment

6.2.1 Introduction

Curing spastic bladder dysfunction caused by sacral spinal cord injuries as describe above can reduce complications in the urinary system, thereby improving patients' quality of life. Upon considering the anatomical features of the sacral nerve root, we designed an experiment to test highly selective sacral root rhizotomy, expecting to improve bladder function by blocking part of the neural pathways to reduce the opportunities for malignant invasion. This experiment also examined the treatment mechanism of selective sacral nerve root rhizotomy and laid a theoretical foundation for the clinical application of this surgical method.

6.2.2 Materials and Methods

6.2.2.1 Preparation of the Animal Model

Twelve male dogs, each weighing 10–12 kg, were prepared into models of complete high-level spinal cord injury. Spastic bladder after spinal cord injury is characterized by detrusor hyperreflexia, contracture, reduction of bladder capacity and decrease of urine storage function, which, together with cramping of the urethral sphincter in most cases, lead to the rise of urethral pressure and increased urethral meatus resistance, which not only have an adverse effect

C. Hou (ed.), *Functional Bladder Reconstruction Following Spinal Cord Injury via Neural Approaches*,
DOI 10.1007/978-94-007-7766-8_6, © Springer Science+Business Media Dordrecht 2014

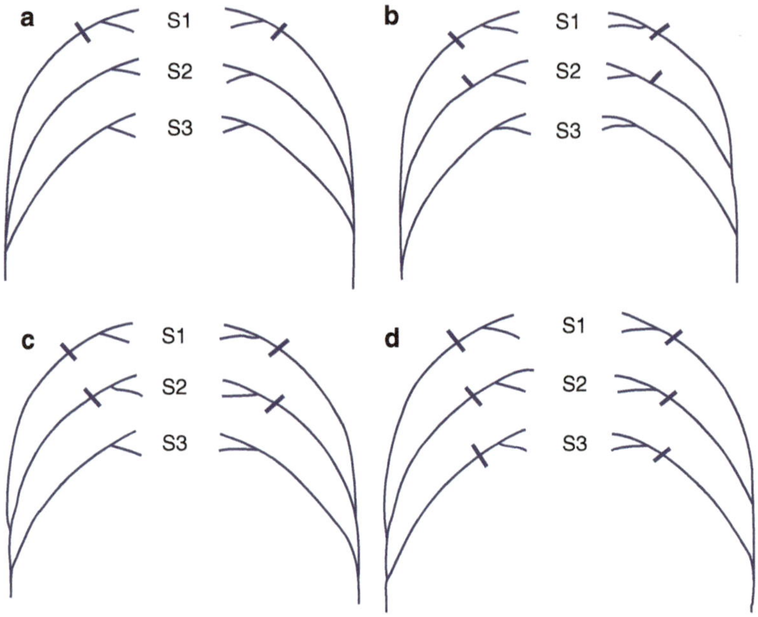

Fig. 6.1 Schematic diagram of selective sacral rhizotomy (**a**) group A; (**b**) group B; (**c**) group C; (**d**) group D

on micturition but also cause the rise of bladder pressure, leading to urine reflux and kidney damage. The study of sacral nerves' control of the lower urinary tract indicates that there is a division of labor of the sacral nerves over different organs and organization of the lower urinary tract, and crossover control exists as well. The S_2 mainly controls the detrusor and partially controls the sphincter. The S_3 mainly controls the sphincter and partially controls the detrusor. The S_4 mainly dominates the penis, and partially controls the detrusor and sphincter. There is a compensatory effect between each of these functions. Therefore, cutting off the S_2 anterior nerve root can theoretically block a major neural signal of the detrusor and keeping the S_4 anterior root can, to a greater extent, avoid the loss of erectile function. In order to achieve the maximum ease of detrusor convulsion while retaining micturition and sexual function, we explored the methods of selective sacral rhizotomy through cutting off the S_{2-4} anterior nerve roots in sequence. Accordingly, this study adopted four different rhizotomy modes and is divided into four groups marked A, B, C and D. In Group A, we selectively cut off the S_2 ventral root and retained the S_3 and S_4; in Group B, we selectively cut off the S_2 ventral root and half of the S_3 ventral root; in Group C, we selectively cut off the S_2 and S_3 ventral roots, retaining the S_4. In Group D, we completely cut off the S_{2-4} nerve roots (Fig. 6.1).

6.2.2.2 Operation Procedure

Proceed with intravenous anesthesia with 25 mg/kg 3 % barbitone sodium. Cut off the spinal cord of a dog at incision T8, and wait 10–20 days for the spastic bladder model to form (Fig. 6.2). Thereafter, anesthetize the dog with the same method as above. Then cut it open in the middle of the hypogastrium and put a No. 12 catheter in the bladder through the incision in the posterior urethra. Penetrate the receiving electrode of the Cantata 2000 electromyographic instrument (Dantec, Denmark) into the detrusor and sphincter's sutured incision. The catheter is connected to a urodynamic pressure gauge, which measures the bladder capacity and urethral pressure. Then cut the L_3–S_4 nerve roots in the middle, resect the vertebral lamina on both sides so that the bilateral S_{2-4} nerve roots are revealed, and locate the nerve root anatomically (Fig. 6.3). Cut open the dural sac and carefully separate the sacral anterior root. Use an insulating rubber sheet to isolate the S_{2-4} nerve roots on both sides from the surrounding tissue to prevent bias caused by electrical stimulation.

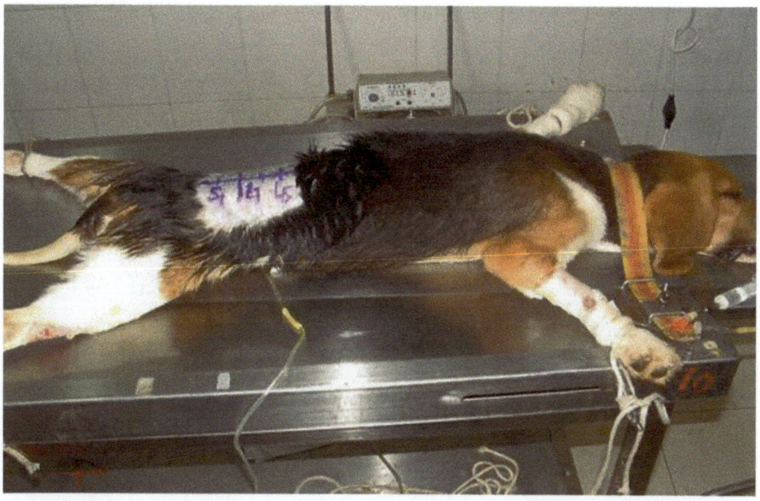

Fig. 6.2 Cut off the spinal cord of a dog at incision T8 for the spastic bladder model to form

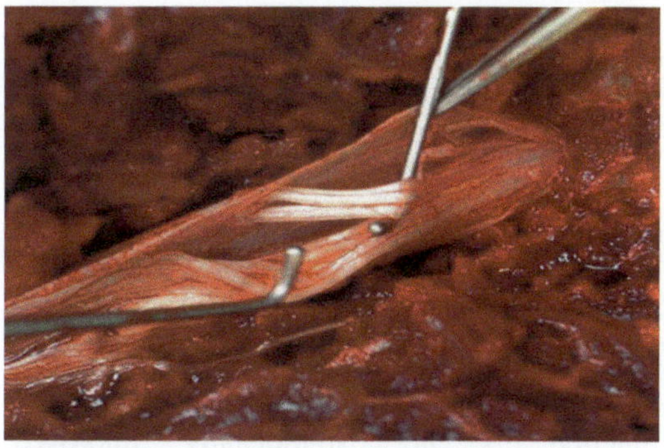

Fig. 6.3 Locate the nerve root of the dogs anatomically

6.2.2.3 Experimental Test Method

Before cutting off the nerve root, record urine flow mechanics and electrophysiology data from which will serve as the control. Selective cutting of the S_2 anterior nerve root is marked as Group A, and carry out urodynamic testing to determine the urinary bladder capacity and urethral pressure. Stimulate the retained sacral nerve root with the single-phase square wave (115 mV, 1.0 ms) electric current of the same strength as that used in the control group. Use the Cantata 2000 electromyographic instrument to record the evoked potentials of the detrusor and sphincter's motor nerve, and calculate the average of the wave crests. Stimulate the retained sacral nerve root with an electric current (1,200 mV, 10 Hz, 25 ms) of the same strength, and observe and record the bladder emptying and penile erectile function. Thereafter, sequentially cut off half of the S_3 anterior nerve root, the rest of the S_3 anterior nerve root and the S_4 anterior nerve root. Conduct the same tests on Group B, Group C and Group D as those on Group A.

6.2.2.4 Statistics

Use a paired *t* test to analyze the urodynamic measured value and changes in electric potential wave crests evoked by muscular movement.

6.2.3 Results

6.2.3.1 Urodynamic Test

As more of the sacral nerve root was cut, bladder capacity increased gradually, bladder compliance obviously improved and urethral pressure was gradually reduced (Table 6.1). For these tests, the measured value of Group B was close to normal (180 ± 30 ml and 60 ± 10 cm H_2O) and there was a significant difference between each group's results ($p<0.05$).

When the retained sacral nerve root is stimulated with a single-phase square wave of the same intensity, there was no obvious difference in the detrusor-motion-evoked electric potential wave crest between the control group and Groups A and B (Table 6.2) ($P>0.05$), while results in Groups C and D obviously decreased and

Table 6.1 Bladder capacity and urethral pressure test result (mean ± standard deviation)

	Control group	Group A	Group B	Group C	Group D
Bladder capacity (ml)	50 ± 20	150 ± 50	180 ± 30	230 ± 50	400 ± 40
Urethral pressure (cm H_2O)	90 ± 20	80 ± 10	60 ± 10	50 ± 10	30 ± 10

Table 6.2 Detrusor sphincter peak potential, examination results (mean ± standard deviation) (mV)

	Control group	Group A	Group B	Group C	Group D
Detrusor	11.5 ± 3.2	11.3 ± 2.62*	11.2 ± 2.52*	8.6 ± 1.45	3.5 ± 1.23
Sphincter	11.2 ± 2.3	10.5 ± 2.78	8.32 ± 2.34	6.5 ± 3.14	3.13 ± 1.34

*$P > 0.05$. Compared with control group

Table 6.3 Voiding of the bladder and erectile function (mean ± standard deviation) (ml)

	Control group	Group A	Group B	Group C	Group D
Bladder voiding volume	30 ± 20	130 ± 30*	180 ± 50*	100 ± 50*	50 ± 30
Erectile function	Existence	Existence	Existence	Abate	Loss

*$P < 0.05$. Compared with control group

exhibited remarkable differences from the other three groups.

6.2.3.2 Urination and Erectile Function Test

With the electric current of same intensity stimulating the sacral nerve root of the sample from each group, the bladder voiding volume result (Table 6.3) illustrates that appropriately cutting off a certain number of sacral nerve roots can significantly increase the bladder urine output, while cutting off too much significantly reduces the urine output. The Group B urine volume increased significantly. Differences in micturition volume between those groups were significant ($p < 0.05$). Penile erectile function was lost by 1 dog each and maintained in the rest of the dogs in Groups A and B, reduced or weakened in Group C, and completely lost in Group D.

6.2.4 Discussion

Treatment of bladder dysfunction caused by spinal cord injury has many years of history [3, 4]. In recent decades, kidney damage and other complications caused by spastic bladder are more clearly understood, and a variety of clinical treatment methods are used to improve patients' urine storage and micturition function and avoid complications. In 1953, Brendler [5] used complete sacral rhizotomy in the treatment of spastic bladder caused by spinal cord injury. By cutting off the sacral reflex arc, he converted an upper motor neuron lesion (UMNL) to a lower motor neuron lesion (LMNL). Although this method is effective in relieving the detrusor sphincter spasm and improving bladder storage, the detrusor sphincter is excessively relaxed and the voiding function is not significantly improved. Meanwhile, the risks of urinary retention and urinary tract infection are increased. In addition, this method can damage the penis erection reflex, leading to impairment of sexual function. In 1968, Manfredi [6] adopted selective posterior sacral rhizotomy. Although the urine storage and micturition function is improved in the short term after the operation, the follow-up observation results indicated that in the long run there is no effective improvement to the bladder function of the majority of patients. Therefore, the method has not been widely adopted. The basic theory of selective sacral rhizotomy is different from those before. With the success of this experimental research, we hope that we can lay a good foundation for clinical application of this method.

The theoretical basis of selective sacral rhizotomy is that the mechanism of bladder detrusor and urethral sphincter spasms after suprasacral cord injury is the partial or total loss of control between advanced central nerves and the spinal nerve system, leading to γ motoneuron excitability, increase of muscle spindle sensitivity, and, subsequently, hyperreflexia and loss of voluntary movement [7–9]. A neural operation to reduce the transfer of malignant signals from voluntary muscle, smooth muscle and muscle spindle γ afferent nerve fibers is the theoretical basis of muscle spasm treatment. However, there is not yet an effective method to distinguish γ afferent nerve fibers from peripheral nerves. Therefore, it is almost impossible to selectively cut off γ afferent nerve fibers, and whether there is an alternative method is the key to the surgical treatment [10–12]. Current research suggests that afferent fibers of γ motor neurons and sacral as well as peripheral nerves may have the same distribution characteristics in functional anatomy, and the nerve fibers controlling the detrusor and sphincter both reach the dominance region through the sacral anterior root. In this regard, selective neurotomy of the sacral anterior root can effectively reduce the malignant nerve signals [13–15].

The experimental results showed that: (1) postoperative bladder capacity and urethral pressure in each group are significantly higher and lower, respectively, than in the control group, and the effect is closely related to the number of sacral nerve roots cut. The bladder capacity and urethral pressure in Group B are both within the normal range. Tip: properly cutting off a certain number of sacral nerve roots can evidently relieve the spasm of the detrusor and sphincter, effectively improve bladder storage function and reduce resistance in the urethral meatus. (2) The results in Groups A, B and C were significantly better than in the control group. Postoperative contrast of bladder capacity and urine output showed that the urine output of Group B increased most significantly without an obvious increase in residual urine. The urine output of Group C and D reduced gradually. Tips: (1) properly cutting off a certain number of sacral nerve roots does not significantly affect the detrusor contraction function. In addition, because bladder capacity increases and the urethral meatus resistance is reduced, the bladder voiding and emptying ability can be effectively improved. (2) Although cutting off too much of the sacral nerve roots can increase the bladder capacity, it may excessively reduce the muscle tension and shrinkage ability of the detrusor, reducing the micturition function of the bladder. (3) The testing results on the electric potential evoked by detrusor motion showed the following curve and peak values of the control group and Groups A and B (Table 6.2). Statistics processing indicated no significant difference. Description: Cutting off an appropriate number of sacral nerve roots has no obvious influence on detrusor contraction, which is consistent with urodynamic testing results. Tips: there is a mechanism of mutual compensation between the sacral nerve root and the sacral plexus. (4) Observations of penile erectile function showed that it is closely related to the S3 nerve, but there are differences among different individuals.

The conclusion of this experimental research: Selective sacral rhizotomy has an obvious curative effect on spastic bladder caused by spinal cord injury, but the cutting site, method and quantity are extremely important to the improvement of bladder storage and micturition function. Cutting too much or too little cannot achieve the desired results and may damage other functions. The treatment used in Group B, i.e., selectively cutting off the S_2 and half the S_3 nerve roots, has the most satisfactory effect.

The experimental results have led the authors to several conclusions: (1) Afferent fibers of γ motor neurons and sacral nerve roots may have the same distribution characteristics in functional anatomy; (2) selective sacral rhizotomy can effectively control γ motoneuron excitability, reduce malignant nerve signals, and properly adjust nerve impulses and muscle tension; (3) partially cutting off the motor nerve also, to some extent, does good to muscle tension adjustment; and (4) there is a mechanism of mutual compensation between the sacral nerve root and the sacral plexus.

The success of this experiment figures to be a new clinical treatment of spastic bladder. The method has the following characteristics: (1) the treatment of spasm of the detrusor and urethral sphincter is highly selective and targeted; (2) this method can effectively avoid damage to the nerves of the penis so as to protect patients' sexual function; (3) because the operation is relatively simple and causes very little damage, it can readily come into wide use.

6.3 Clinical Application

6.3.1 Introduction

To practically solve spastic bladder caused by spinal cord injury, avoid damage to renal function and cultivate self-discipline of the bladder, we performed selective sacral rhizotomy on SCI patients.

6.3.2 Indications

Patients who presented with complete suprasacral SCI with neurogenic bladder and had stable neurological injury and no contraindications for general anesthesia or surgery.

6.3.3 Surgical Technique

Using the conventional lumbosacral vertebrae ($L_5 \sim S_1$) posterior median incision, cut open the subcutaneous tissue and supraspinal ligament. Separate muscle on both sides of the spinous process, reveal the bilateral vertebral plate on both sides of the lumbosacral vertebrae, and conduct anatomical localization of the lumbosacral vertebrae. Perform bilateral laminectomy on $L_5 \sim S_1$, revealing the dural sac and extracapsular nerve root. Setting the $L_5 \sim S_1$ intervertebral space as the mark, locate the extracapsular S_1 nerve root and then open the dural sac to reveal the cauda equina. Locate the intracystic S_1 nerve root according to the extracapsular S_1, and then locate the S_{2-4} nerve roots one after another. Distinguish and separate the anterior and posterior roots according to their anatomical characteristics. Generally, the anterior root is located on the ventral side and is relatively thinner, while the posterior root is on the dorsal side and thicker. Stimulating the S_{2-4} anterior nerve roots in sequence with the Cantata 2000 electrical stimulation instrument at the same intensity (20 mV, 30 Hz, 5–10 s), observe the shrinkage reaction of the bladder detrusor and urethral sphincter, identify the most sensitive nerve root and cut it off.

Application of a self-made pressure capsule in the operation can help to judge the vasomotor state of the urinary bladder detrusor muscle directly so as to further distinguish the sacral nerve innervation condition. At the same time, penetrate the receiving electrode of the Cantata 2000 type electromyographic instrument directly into the anal sphincter to observe the reaction of the urethral sphincter; the urethral and anal sphincters are both innervated by the pudendal nerve.

6.3.4 Discussion

Highly selective sacral rhizotomy has a distinct clinical curative effect [16]. We think that, as a new surgical method, highly selective sacral anterior root rhizotomy not only avoids the drawback of complete sacral rhizotomy, namely, that the spastic bladder turns into a flaccid bladder, but is also superior to the poor long-term outcome of selective posterior rhizotomy. Meanwhile, the operation provides more complete protection of patients' erectile function. This new surgical method has the following features: (1) treatment to the detrusor of bladder, urethral sphincter spasm has high selectivity and pertinency; (2) the method can effectively avoid damage to nerves governing the penis, thus protecting the erectile function; (3) the operation is simple with little injury, and therefore easy to spread.

Various urodynamic testing devices produced domestically and abroad have been developed to a very advanced level with automatic detection, recording and many other functions. However, most of these devices are voluminous, heavy and are immobile. Patients can be examined only in a fixed place. Since it is not easy for paraplegic patients to be moved, it is very difficult or even impossible to perform pressure examination on them with the above devices. To solve this problem, we designed and manufactured a simple three-lead urodynamic manometer, which has been applied satisfactorily to intraoperative and ward monitoring.

The simple three-lead urodynamic manometer we designed may simultaneously measure the bladder, urethra and rectal pressure. An electromyography test can also be conducted at the same time if necessary. Therefore, measurement of lower urinary tract function after spinal cord injury can be performed more comprehensively and systematically. Since the simple manometer is very small, it is easy to carry and its use is not limited by time and location. Therefore, at any time the device can carry out pressure test on paraplegic patients who are not easy to move, solving the problem that urodynamic testing cannot be systematically conducted on such patients in a timely manner. At the same time, the device is direct, precise and easy to operate during surgery. In addition, the device can be placed before a patient's bed. The patient can directly observe pressure change, based on which he or she can consciously train his or her own detrusor systolic function, laying a good foundation for a self-disciplined bladder. Since the manometer is relatively simple in structure and easy to manufacture, it is also very suitable for basic-level hospitals.

References

1. Liu MX. Evaluation and classification of bladder function after spinal cord injury. Chin J Spine Spinal Cord. 1998;8:10–1.
2. Liu MX. Experimental study of sacral nerve distribution and its influence on lower urinary tract function. Acad J Second Mil Med Univ. 1999;20:298–9.
3. Hackle RH. A 25-year prospective mortality study in the spinal lord injured patient: comparison with the long-term living paraplegic. J Urol. 1977;117: 486–8.
4. Rijkhoff NJ, Holsheimer J, Koldewijn EL. Selective stimulation of sacral nerve roots for bladder control: a study by computer modeling. IEEE Trans Rehab Eng. 1994;41:413–24.
5. Brendler H, Krueger EG, Lerman P, et al. Spinal root section in treatment of advanced paraplegic bladder. J Urol. 1953;70(2):223–9.
6. Manfredi RA, Leal JF. Selective sacral rhizotomy for the spastic bladder syndrome in patients with spinal cord injuries. J Urol. 1968;100(1):17–20.
7. Jonas U, Tanagho EA. Studies on the feasibility of urinary bladder evacuation by direct spinal cord stimulation. II. Post stimulus voiding: a way to overcome outflow resistance. Invest Urol. 1975;13: 151–3.
8. Jonas U, Heine JP, Tanagho EA. Studies on the feasibility of urinary bladder evacuation by direct spinal cord stimulation. I. Parameters of most effective stimulation. Invest Urol. 1975;13:142–50.
9. Jonas U, Jones LW, Tanagho EA. Spinal cord versus detrusor stimulation. A comparative study in six acute dogs. Invest Urol. 1975;13:171–4.
10. Hald T, Agrawal G, Kantrowitz A. Studies in stimulation of the bladder and its motor nerves. Surgery. 1966;60:848–56.
11. Timm GW, Bradley WE. Electrostimulation of the urinary detrusor to effect contraction and evacuation. Invest Urol. 1969;6:562–8.
12. Holmquist B. Electromicturition by pelvic nerve stimulation in dogs. Scand J Urol Nephrol. 1968;2:1–27.
13. Ingersoll EH, Jones LL, Hegre ES. Effects on urinary bladder of unilateral stimulation of pelvic nerves in the dog. Am J Physiol. 1957;189:167–70.
14. Habib HN. Experience and recent contributions in sacral nerve stimulation for volding in both human and animal. Br J Urol. 1967;39:73–83.
15. Heine JP, Schmidt RA, Tanagho EA. Intraspinal sacral root stimulation for controlled micturition. Invest Urol. 1977;15:78–82.
16. Brindley GS, Polkey CE, Rushton DN, Cardozo L. Sacral anterior root stimulator for bladder control in paraplegia: the first 50 cases. J Neurol Neurosurg Psychiatry. 1986;49:1004–14.

7 Reconstruction of Bladder Innervation Below the Level of Spinal Cord Injury for Inducing Urination by Achilles Tendon-to-Bladder Reflex Contractions

Haodong Lin, Hongbin Zhong, and Chunlin Hou

7.1 Background

Neuropathic bladder caused by spinal cord injury (SCI) has been a challenge for researchers and clinicians for a long time. Most treatments focus on improving micturition, because there is no effective treatment for the spinal cord injury. The most common treatments for improving micturition are magnetic stimulation [1, 2], reconstruction of the detrusor with neuromuscular flaps [3, 4], and selective excision of the sacral nerve root [5]. However, satisfactory outcomes have been reported only for a few of these methods. Currently, the sacral root electrostimulation technique developed by Brindley is the only known effective clinical treatment, but the results with this treatment are inconsistent. Recently, surgical reconstruction of bladder innervation has been undertaken in paraplegic patients to upgrade their quality of life and to lower the mortality rate.

H. Lin • C. Hou, M.D. (✉)
Department of Orthopedic Surgery, Changzheng Hospital, The Second Military Medical University, Fengyang Road 415, Shanghai 200003, People's Republic of China
e-mail: haodonglin@hotmail.com; chunlin_hou@163.com

H. Zhong
Department of Orthopedic Surgery, The First Affiliated Hospital of CPLA General Hospital, Beijing, People's Republic of China
e-mail: zhonghb304@yahoo.cn

7.2 Animal Experiment

7.2.1 Introduction

Previous studies on bladder function reconstruction employed the normal somatic reflex above the paraplegic level for reinnervation. For example, in 1991, Chuang et al. [6] established an intradural anastomosis from the spinal nerve ventral root above the SCI level to the sacral nerve ventral root below the SCI level in a rat model, and successfully repaired bladder dysfunction after low level SCIs. In 1994, Xiao and Godec successfully established an artificial skin-CNS-bladder reflex pathway in a rat model of L5 paraplegia by creating an intradural anastomosis from the central end of the L4 spinal nerve ventral root to the peripheral end of the bladder-governing L6 spinal nerve ventral root [7]. In both these studies, the normal somatic reflex above the paraplegic level was utilized to reconstruct bladder function. However, we think that it would be more feasible to adopt the patellar or Achilles tendon reflexes below the paraplegic level in the case of SCIs sustained above the medullary cone, since the lower central circuit remains intact in such injuries. To test this hypothesis, we performed a series of experiments on the patellar tendon-bladder reflex arc.

C. Hou (ed.), *Functional Bladder Reconstruction Following Spinal Cord Injury via Neural Approaches*,
DOI 10.1007/978-94-007-7766-8_7, © Springer Science+Business Media Dordrecht 2014

7.2.2 Materials and Methods

7.2.2.1 Animals and Procedure

The sample consisted of 6 beagle dogs (weight, 10–12 kg). Dogs 1–4 were observed for 6-month short-term postoperative outcomes, and dogs 5 and 6 were examined for 18-month long-term postoperative outcomes. Initially, all dogs were anesthetized intravenously with 3 % sodium pentobarbital (25 mg/kg). The knee-tendon reflex center of the dog was located in the area from the L4 to L6 spinal segment, and the spinal nerve center governing the bladder was located along the S1 to S3 segment; a posterior median incision was then made from L3 to S2. The right and left laminas were removed, and the dura was incised to expose the cauda equina. The nerve root was localized by manual observation, and their location was confirmed by electrostimulation. Two anterior motor roots, the right S2 and the right L5, were cut off at the same level. Then, the anastomosis of the distal end of the right S2 anterior motor root and the proximal end of the right L5 anterior motor root were employed by utilising 11–0 monofilament nylon under a 10-power microscope. The right L5 posterior sensory root remained intact (Fig. 7.1). Finally, the anastomosis was labeled with a silk thread, in order to protect the nerve root at the time of postoperative examinations, and absorbable sutures were used to stitch up the internal wounds. The intact left S2 root was deployed as the normal control.

7.2.2.2 Electrophysiologic Examination

The dogs were examined under anesthesia at the set postoperative times (6 months and 18 months). In the case of dogs 1–4, the skin was incised along the original incision. The scar outside the dura was stripped off. The dura was incised to separate the cauda equina carefully. The anastomosis was identified by means of the silk thread that was used as a label. For the experimental measurements, the recording electrode was inserted at the distal end of the anastomosis in the right S2 anterior motor root. Meanwhile, the stimulation electrode was inserted at the right L5 posterior sensory root. Both electrodes were connected to the EDJ-V Biosignal system (Shanghai Teaching Equipment Co., Shanghai, China), and unidirectional rectangular pulses of 115 mV were delivered for 1.0 ms. Then, action potential (AP) curves were recorded. For the control measurements, the recording electrode and the stimulation electrode were placed at the left S2 anterior motor root and the left S2 posterior sensory root respectively. The stimulation parameters were the same as those for the experimental measurements.

In the case of dogs 5 and 6, a median incision was made in the lower abdomen. For experimental measurements, the electrodes were inserted into the upper anterior wall of the bladder detrusor muscle, in order to stimulate the right L5 posterior sensory root, the right femoral nerve, and the afferent nerve of the knee-tendon reflex. The electrodes were connected to the Cantata 2000 muscular electrograph (Dantec Medical, Copenhagen, Denmark), and pulses of 3.8 mA were delivered at 1.0 Hz. Electromyograms (EMGs) of the detrusor muscle were recorded. For control measurements, the left S2 posterior sensory root was stimulated with the same parameters as those for the experimental measurements.

In both groups of dogs, paraplegia was induced by transecting the spinal cord acutely at the T10 canal level. Observations were carried out 48 h after paraplegia.

7.2.2.3 Measurement of Intravesical Pressure

As mentioned before, 48 h after the paraplegia model was induced, the dogs were observed under anesthesia. A median incision was made in the lower abdomen to unveil the bladder cervix and posterior urethra after anesthesia. A 12-F urinary catheter was inserted into the bladder through a small incision on the posterior urethra, after which the incision was sutured. The urinary catheter was then linked to a self-designed urodynamic 3-lead simplified pressure machine [8]. Through the fluid inlet, it was approximately 150 mL of normal saline injected into the bladder. In order to facilitate observing the results, the water column of the pressure test tube was maintained in the baseline of 10-cm. The fluid inlet

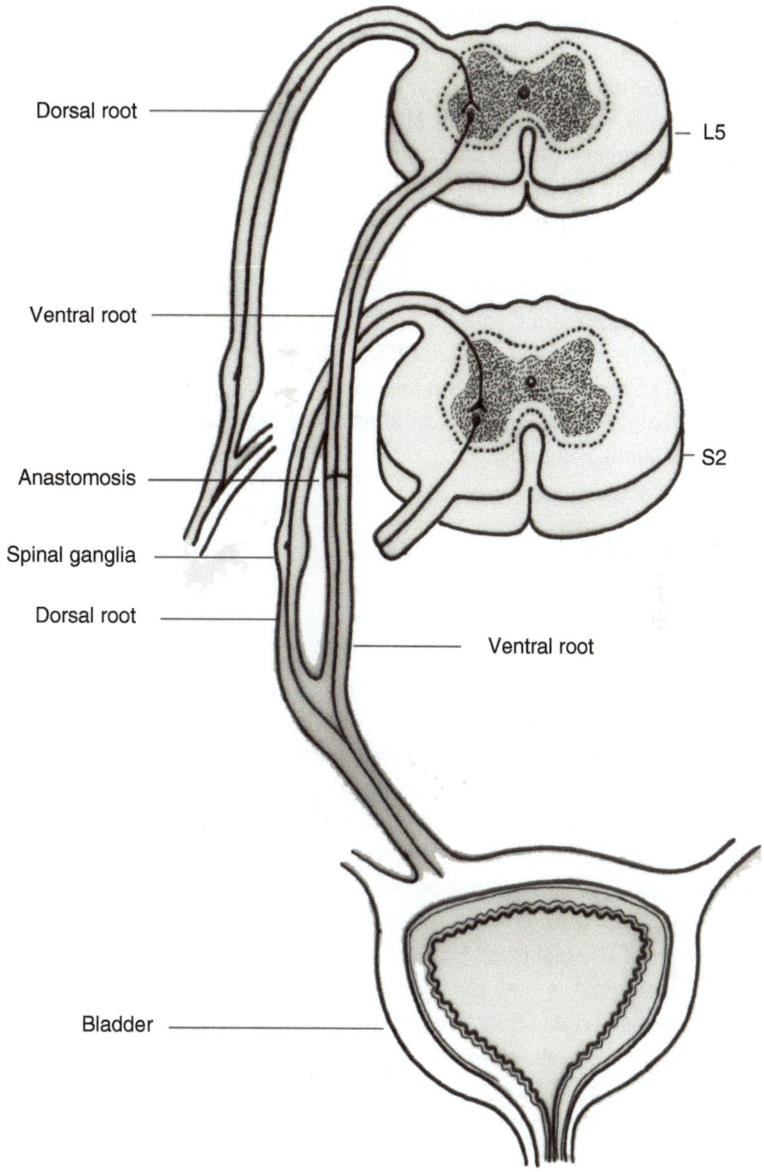

Fig. 7.1 Illustration of the tendon-spinal cord-bladder artificial reflex arc in canines

was turned off. After this, the cauda equina was exposed via the original median approach to look for the previous anastomosis of two anterior motor roots (i.e. the right L5 and S2) in all dog samples. In the cases of Dogs 5 and 6, a right femoral anterior lateral incision was conducted to expose the femoral nerve acted as the afferent nerve of the knee-tendon reflex.

A series of stimulation currents of 1,000 mV at 10 Hz were delivered for 2 s from an EJD-V Biosignal system, in order to stimulate the proximal end of the right L5 anterior motor root anastomosis, right L5 posterior sensory root, left S2 anterior motor root, left L5 posterior sensory root, and right femoral nerve. A medical percussion hammer was used to percuss the right knee tendon and the left knee tendon for 10 s at a frequency of 2 times per second. Increase in bladder pressure (cm H_2O) and duration of urination (T) were recorded. The left S2 anterior motor root served as the normal control, while the left L5 posterior sensory root and left knee tendon were utilized as negative controls.

7.2.2.4 Neurohistologic Observations

A nerve specimen was derived from the region 0.5 cm proximal and 1.0 cm distal to the anastomosis, and it was stained with HE in order to count the number of fibers containing medullary nerves with the Leica FW4000 Image Analysis System (Leica, Solms, Germany). The passing rate of myelinic nerve fibers was determined by dividing the number of fibers at the distal end with that at the proximal end.

Data were analyzed using SPSS12.0 and expressed as mean ± SD. The Student's *t*-test was used to analyze data. $P<0.05$ was considered to indicate statistical significance.

7.2.3 Results

7.2.3.1 Electrophysiologic Examination

The same current was released to stimulate L5 posterior sensory nerve root before paraplegia, which was repeated 48 h after complete paraplegia at the T10 canal level. In all 6 dogs, AP curves were found in S2 anterior motor root distal to the anastomosis. The morphology and amplitude of these curves were similar to those of the control side (Fig. 7.2). After the 48 h of complete paraplegia at the T10 canal level, EMG of the detrusor muscle was found in both the right L5 posterior sensory root and right femoral nerve. They shared similarity in both morphology and amplitude with those of the left S2 posterior sensory root of the control, whereas stimulation of the left L5 posterior sensory root failed to trigger an EMG response of the detrusor muscle (Fig. 7.3).

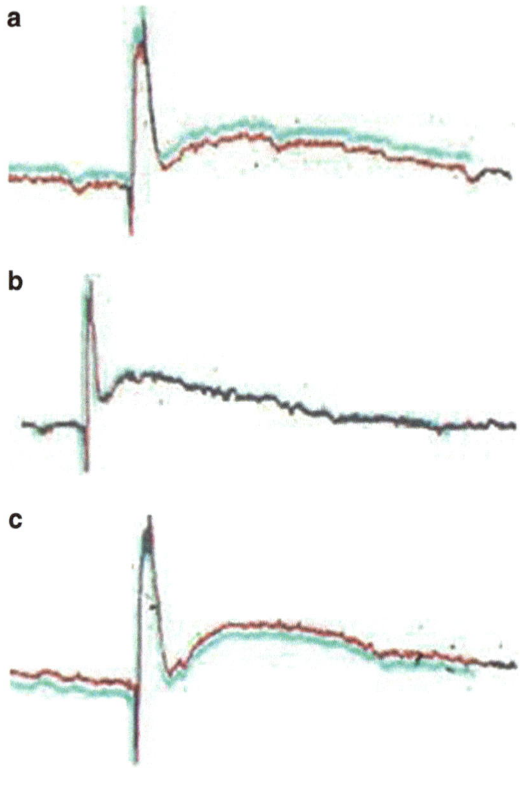

Fig. 7.2 AP curves of dogs 5 (*left*) and 6 (*right*), which were obtained by stimulation at the left S2 anterior motor root and recording at the left S2 posterior sensory root (**a**). The right L5 posterior sensory root was stimulated before (**b**) and 48 h after complete paraplegia (**c**), with the nerve recordings taken at the distal ends of the anastomosis

7.2.3.2 Intravesical Pressure Measurement

In dogs 1–4, mean bladder contraction and mean duration of the urination were 59 % ± 23 % and 83 % ± 31 % that of the control group; mean bladder contraction and mean duration induced by percussion of the right knee tendon were 38 % ± 27 % and 51 % ± 37 % that of the control group respectively. In dogs 5 and 6, mean bladder contraction and mean duration were 84 % ± 5 % and 88 % ± 6 % that of the control group; the values obtained on percussion of the right knee tendon were 62 % ± 5 % and 84 % ± 12 % that of the control group. Thus, compared with the 6-month group, the 18-month group experienced a significant increase in bladder pressure ($P<0.05$). Moreover, the duration of urination was also significantly longer in the 18-month group ($P<0.05$). However, no significant different in duration between the groups was observed with percussion of the right knee tendon ($P>0.05$).

7.2.3.3 Neurohistologic Observations

HE staining unveiled that numerous nerve fiber growths passed through the anastomoses on the experimental side. The longitudinal section had regenerated nerve fibers that were well arranged and grew in the same direction. The cross-sectional

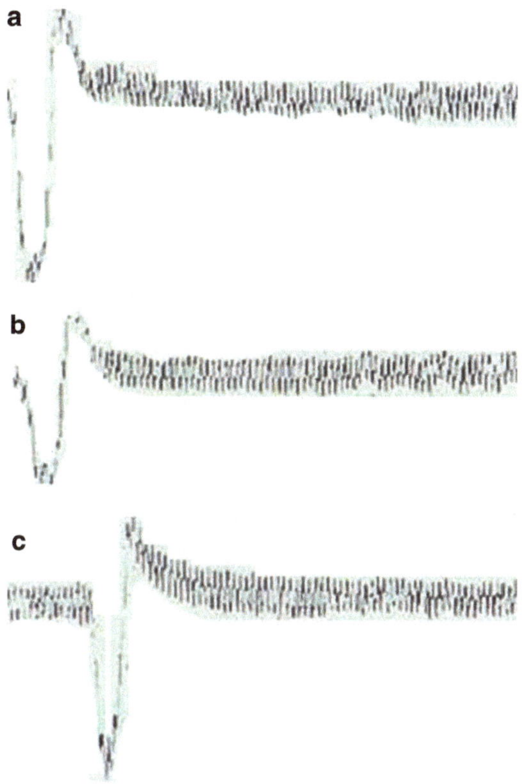

Fig. 7.3 EMGs of the detrusor muscle of dogs 5 (*left*) and 6 (*right*), which were obtained by stimulation at the left S2 posterior sensory root (**a**), right L5 posterior sensory root (**b**), and right femoral nerve (**c**), which was the afferent nerve of the knee-tendon reflex

view showed regenerated nerve fibers that presented with a typical myelinated nerve fiber structure (i.e., myelin sheath at the periphery and nerve axon in the center). The figures for nerve fibers distal and proximal to the anastomosis on the experimental side were 574±261 and 988±124 respectively. The passing rate of myelinated nerve fibers was in the range of 58.1 %±13.4 %. By contrast, the figure for fibers of the left S2 anterior root was around 867±349. The difference between the two sides was statistically significant ($P<0.05$).

7.2.4 Discussion

The purpose of this study was to explore the potential effects of creating a new knee tendon-bladder reflex arc below the paraplegic level in dogs that had sustained SCI above the medullary cone. A combination of electrophysiological and neurohistological methods were used to study the new reflex arc established. Specifically, electrical stimulation in the six dogs showed that the new reflex arc was able to induce bladder contraction; in two dogs however, bladder contraction was also induced by percussion of the knee-tendon. This can be attributed to poor quality of the anastomosis without enough axons passing through the stoma in a timely fashion. It may have also resulted from mismatch of the nerve fibers.

We found that bladder contraction in the 18th month was significantly stronger than that in the 6th month after surgery; this indicated that long-term axon regeneration, reinstitution of the bladder nerve, and recovery of bladder function were better than those in the early stages after the operation.

Our neurohistologic results showed that numerous nerve fiber growths passed through the anastomosis at the experimental sides. This also indicated that the autonomic nerve fibers of S2 can be reinnervated by the somatic nerve fibers of L5. Therefore, reconstructed bladder reinnervation below the level of SCI could facilitate urination via the knee tendon. Moreover, the motor impulses of the somatic reflex caused by percussion of the knee tendon were transmitted to the bladder through the motor efferent branch, leading to spontaneous contraction of the bladder.

Our anastomosis site differs markedly from those reported by Xiao and Godec [7] and Chuang et al. [6], who used the normal somatic reflex above the injury level for reinnervation of the bladder. We used a healthy tendon reflex below the injury level to reconstruct an artificial bladder reflex. We believe that this new approach provides theoretical references for clinical applications.

7.3 Clinical Application

7.3.1 Introduction

Patients with complete SCI above the conus medullaris (spinal micturition center) are prone to spastic bladder. These patients present with progressive urological dysfunction characterized by involuntary bladder contractions, reduced volume capacities, leakage around the urethral catheters, recurrent infections, vesicoureteral reflux, and hydronephrosis, which lead to atrophy of the kidney, uremia, and eventually death. The cauda equina sacral nerve root approach is the main neurological procedure used to reconstruct bladder function. Animal experiments have demonstrated the possibility of using somatic reflexes to reconstruct bladder function and control urination [9–12]. Based on the somatic nerve–autonomic nerve artificial bladder reflex arc reported by Xiao and colleagues [9, 13], we used an existing healthy somatic reflex and constructed a patellar tendon–spinal cord–bladder artificial reflex arc via nerve anastomosis in dogs. In this arc, motor impulses of the somatic reflex produced by percussion of the patellar tendon are transmitted to the bladder via the motor efferent branch, inducing spontaneous contraction of the bladder [14]. The results of electrophysiological studies, vesicular pressure measurement, and HRP tracing indicated that this new reflex pathway was successful and effective. The somatic motor nerve fibers of the L-5 ventral root can grow into the myelin sheath of the parasympathetic nerve fibers of the S-2 ventral root through the anastomosis, creating a new nonphysiological nerve connection with the target organ, the detrusor urinae. These findings provided theoretical references for clinical use of the approach. Based on the results of these preclinical experiments, we attempted to reconstruct bladder innervation below the level of SCI in order to induce urination by Achilles tendon–bladder reflex contractions in paraplegic patients.

7.3.2 Indications

Patients who presented with complete suprasacral SCI with neurogenic bladder and had stable neurological injury and no contraindications for general anesthesia or surgery.

7.3.3 Surgical Technique

General endotracheal anesthesia was induced in the patients. Then, a 3-channel Foley catheter was introduced into the bladder, with one channel connected to a drainage bag, the second connected to a pressure transducer leading to an urodynamic unit, A standard laminectomy from L-5 to S-3 was performed with the patient lying in the prone position. The dura mater was cut open via a paramedian incision, and the dorsal and ventral roots of the S1–4 nerves were exposed. The S-1 nerve root was located by using the L5/S1 intervertebral space as a marker; the S2–4 nerve roots were located in descending order. The ventral and dorsal roots at the dural incision were identified based on their anatomical characteristics: the smaller and darker medioanterior rootlet (usually single) was the ventral component, whereas the larger and paler lateroposterior rootlet (usually double) was the dorsal component.

The ventral roots of S-1, S-2, and S-3 were identified, separated from their respective dorsal roots by microdissection, and tested by electrostimulation to confirm their locations. Electrostimulation of the S-2 and S-3 ventral roots should induce bladder contraction and an increase in intravesical pressure. The nerve root whose intravesical pressure increased the fastest and was the highest was regarded as the dominant root governing the bladder. The unilateral S-1 ventral root and the dominant root at the same level were microsurgically cut and anastomosed with 9–0 sutures (Fig. 7.4). In the study group, anastomosis of the S-1 ventral root and both the S-2 and S-3 ventral roots was performed in the patients to form an Achilles tendon–spinal cord–bladder reflex arc. The wound was sutured in three layers, along with placement of an external drain. After surgery, broad-spectrum antibiotics were given for 3 days. All patients were moved to rehabilitation clinics 2 weeks after surgery.

7.3.4 Case Report

A 36-year-old man was referred to our hospital for evaluation and treatment of neurogenic bladder with complete suprasacral SCI in June 2004. He had fallen from an elevated height, which resulted in a compression fracture of the T7 vertebra and consequently complete paraplegia.

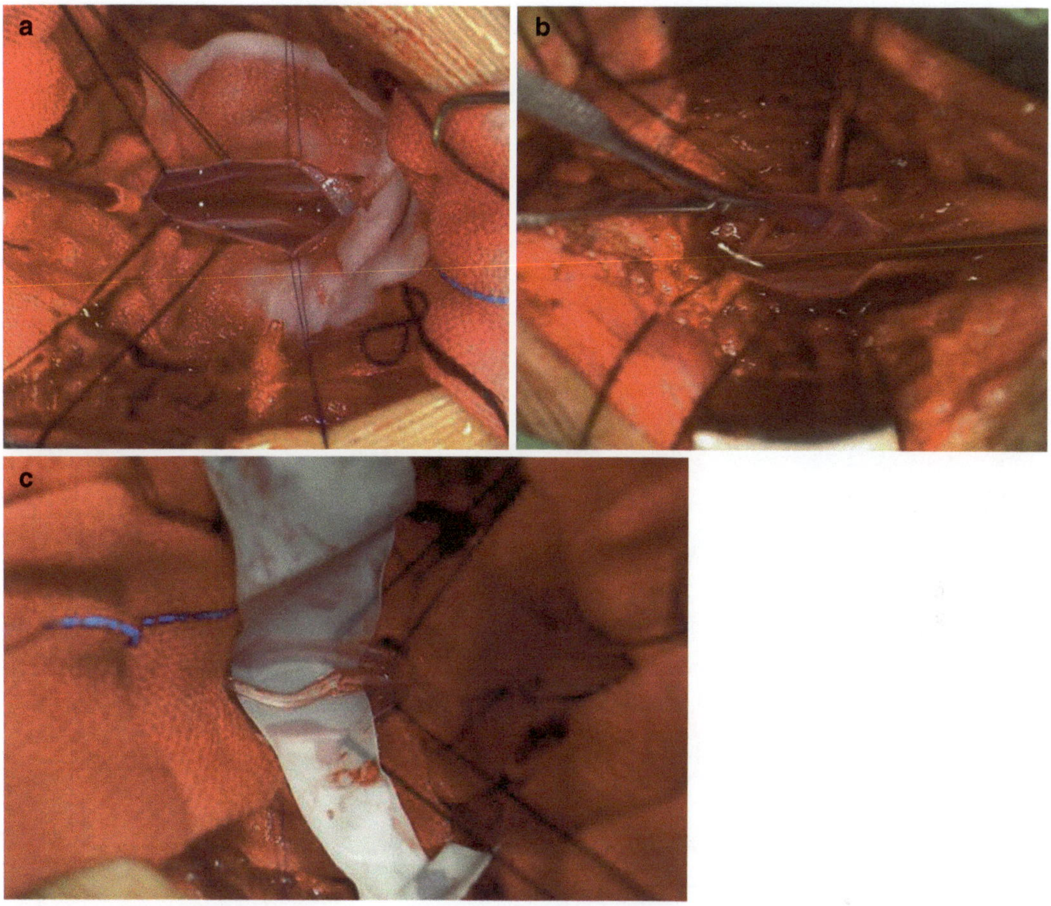

Fig. 7.4 Intraoperative photographs showing the surgical procedure for reconstruction of the bladder reflex pathway in a 20-year-old man. (**a**) Standard laminectomies from L5 to S3 were performed with the patient lying in the prone position. The dura mater was opened up through a paramedian incision, exposing the dorsal and ventral roots of the S1, S2, S3, and S4 nerves. (**b**) The ventral roots of S1, S2, and S3 were identified and separated from their respective dorsal roots by microdissection. (**c**) The S1 and S2/3 ventral roots were anastomosed

The patient carried condom-like urine bags, into which urine drained at 20–60-min intervals during the day. Urine output each time was small (20–120 ml), and urination was frequent (10–15 times/day). As expected, the patient developed a serious urinary infection, for which antibiotic therapy was started. Physical examination showed that he had lost the ability to move his lower extremities; moreover, investigation of the hyperactive patellar reflex showed paralysis with spasms. Positive findings were obtained for ankle clonus and the Babinski sign. Anal and bulbospongiosus reflexes were elicited in the patient, which suggested that medullary cone function was good. Preoperative urodynamic examination suggested detrusor external sphincter dyssynergia. Surgical reconstruction of the bladder reflex pathway was scheduled and performed 4 months after the injury. The unilateral S-1 ventral root and the S-2 and S-3 ventral roots were microsurgically cut and anastomosed with 9–0 sutures (Fig. 7.5).

The patient was followed up for 6 years. The reflex arc was constructed 10 months after the operation. The patient experienced recovery of bladder storage and voiding functions, with no nocturnal urinary incontinence. The Achilles tendon–spinal cord–bladder reflex (percussion of the Achilles tendon) became functional ~12 months after the operation. The patient voided approximately 5–10 s after the beginning of Achilles tendon stimulation. Urodynamic and

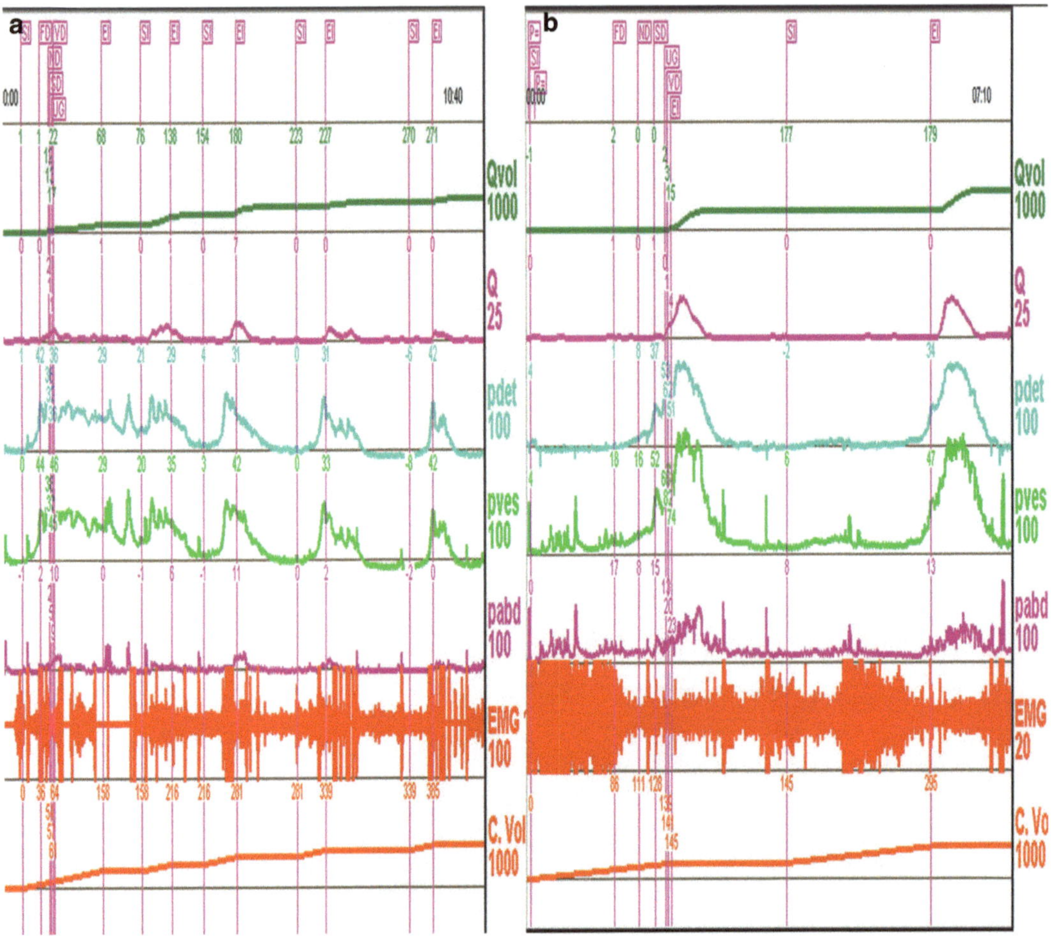

Fig. 7.5 Urine pressure and flow diagrams showing the results of urodynamic tests. (**a**) Preoperative findings showing strong involuntary detrusor contractions, elevated Pdet, and increased electrical activity on the EMG, which were consistent with DESD. (**b**) The findings at 12 months showing generally normalized bladder storage and emptying function. Satisfactory voiding without DESD was initiated by percussion of the Achilles tendon

EMG studies of the external urethral sphincter documented the changes in bladder activity, from hyperreflexia associated with detrusor external sphincter dyssynergia to almost normal findings. Any dysfunction identified preoperatively by EMG was found to have been corrected after the operation. On comparison of the patients' pre- and postoperative urodynamics, we found significant improvement in bladder function at the 12th postoperative month. The average bladder capacity increased from 243 to 356 ml, residual urine volume decreased from 208 to 35 ml, maximum Pdet decreased from 84.2 to 67.0 cm H_2O, and maximum flow increased from 3.4 to 21.6 ml/s. The incidence of urinary tract infections gradually decreased after the 6th month and disappeared as voiding function recovered.

7.3.5 Discussion

On the basis of our preclinical experiments in dogs, we were able to successfully reconstruct bladder innervation below the level of SCI and induce urination by Achilles tendon–bladder reflex contractions in 12 paraplegic patients. In our method, the S-1 was anastomosed with the S-2 and/or S-3 ventral root to reconstruct the bladder reflex arc. This approach resulted in enlarged bladder volume, increased urine output, decreased urinary incontinence, and improved bladder function [15]. In the early postoperative period, the artificial bladder reflex arc was not yet established. Severing the dominant sacral nerve root governing the organ helped correct spasmodic bladder. With the regeneration of nerve roots and reconstruction of the artificial reflex arc in the later stage of the procedure, reinnervation was completed. Furthermore, the Achilles tendon reflex was established as a new trigger point for inducing bladder voiding. The results of our study are similar to the report by Xiao et al. [13] where 15 patients with spasmodic bladder underwent microanastomosis of the proximal end of the L-5 ventral root with the distal end of the S2/3 ventral root in the dural cyst for reconstruction of urinary function through the skin-CNS-bladder reflex arc. Of the 15 patients, 10 (67 %) achieved controlled urination. The difference between this study and ours is that we used the Achilles tendon (S-1) close to the dominant nerve root (S-2 or S-3) instead of the L-5 ventral root. Because the distance is shorter, nerve function can recover in the early stage of the procedure.

The Achilles tendon–spinal cord–bladder reflex arc has the following advantages. Compared with the previous approaches, urine storage is improved by breaking off of the dominant nerve root governing the bladder. Moreover, it was possible to achieve stress-free nerve anastomosis without nerve transplantation because the L5 nerve root overlapped the S2 nerve root in the cauda equina. Additionally, this method is superior to the traditional ones with regard to the nerve regeneration time, which is shorter because the anastomotic stoma is relatively narrow. Furthermore, the nerves below the injury level that were dysfunctional are used to reconstruct voiding with no further functional loss of the lower limbs. Although the unilateral sacral anterior nerve root is severed, another lateral anterior nerve root, the posterior nerve root, and sacral cord are still intact. Thus, bowel and sexual functions are not affected. Finally, anastomosis of simple motor nerves avoids mismatching of axons, so nerve function recovers smoothly. Further studies are required to explore the mechanisms underlying the interactions between the somatic and autonomic motor nerves.

However, this approach still has some disadvantages. Compared with indwelling catheterization and Crede'/Valsalva maneuvers, it is more invasive and therefore carries a higher risk of skin breakdown. Another obvious deficit of this approach is that the constructed artificial reflex pathway is only a new trigger of urination rather than a complete reflex arc. Thus, patients may still lack sensation of urination even after reconstruction of the reflex arc. To some extent, they are unable to void in a timely and flexible manner, which may affect their quality of life. In addition, because the bladder innervation is reconstructed with the tendon reflex, stretching the knee or ankle joint while standing could potentially produce reflex urination.

Thus, while this method is more advantageous than currently used techniques in many ways, there are still some drawbacks that need to be overcome.

References

1. Fujishiro T, Takahashi S, Enomoto H, et al. Magnetic stimulation of the sacral roots for the treatment of urinary frequency and urge incontinence: an investigational study and placebo controlled trial. J Urol. 2002;168:1036–9.
2. Reitz A, Knapp PA, Frey S, et al. Functional magnetic stimulation of the spinal cord: a urodynamic study in healthy humans. Neurourol Urodyn. 2004;23: 148–53.
3. Ninković M, Stenzl A, Hess M, et al. Functional urinary bladder wall substitute using a free innervated latissimus dorsi muscle flap. Plast Reconstr Surg. 1997;100:402–11.
4. Stenzl A, Ninkovic M, Willleit J, et al. Free neurovascular transfer of latissimus dorsi muscle to the bladder: experimental studies. J Urol. 1997;157:1103–8.
5. Gasparini ME, Schmidt RT, Tanagho EA. Selective sacral rhizotomy of the reflex neuropathyic bladder: a report on 17 patients with long-term follow-up. J Urol. 1992;148:1207–10.
6. Chuang DC, Chang PL, Cheng SY. Root reconstruction for bladder reinnervation: an experimental study in rats. Microsurgery. 1991;12:237–45.
7. Xiao CG, Godec CJ. A possible new reflex pathway for micturition after spinal cord injury. Paraplegia. 1994;32:300–7.
8. Liu M, Hou C, Bao J. A new simple uroflometer. Acad J Second Mil Med Univ. 1997;18:392–3.
9. Xiao CG, Du MX, Li B, et al. An artificial somatic-autonomic reflex pathway procedure for bladder control in children with spina bifida. J Urol. 2005;173:2112–6.

10. Wang J, Hou C, Jiang J, et al. Selection of the sacral nerve posterior roots to establish skin-CNS-bladder reflex pathway: an experimental study in rats. Microsurgery. 2007;27:118–24.
11. Sievert KD, Xiao CG, Hennenlotter J, et al. Voluntary micturition after intradural nerve anastomosis. Urologe A. 2005;44:756–61.
12. Vorstman B, Schlossberg S, Kass L. Investigations on urinary bladder reinnervation. Historical perspective and review. Urology. 1987;30:89–96.
13. Xiao CG, Du MX, Dai C, et al. An artificial somatic-central nervous system- autonomic reflex pathway for controllable micturition after spinal cord injury: preliminary results in 15 patients. J Urol. 2003;170: 1237–41.
14. Zheng XY, Hou CL, Zhong HB, et al. Reconstructed bladder innervation below the level of spinal cord injury: the knee-tendon to bladder artificial reflex arc. J Spinal Cord Med. 2009;32:79–85.
15. Lin H, Hou C, Zhen X, Xu Z. Clinical study of reconstructed bladder innervation below the level of spinal cord injury to produce urination by Achilles tendon-to-bladder reflex contractions. J Neurosurg Spine. 2009;10:452–7.

Reconstruction of Bladder Innervation Above the Level of Spinal Cord Injury for Inducing Urination by Abdomen-to-Bladder Reflex Contractions

8

Haodong Lin, Jinwu Wang, Guibin Zhong, and Chunlin Hou

8.1 Background

Spinal-cord injuries (SCIs) can seriously damage somatic motion and sensation. However, they can also cause visceral neurogenic dysfunctions, among which compromised micturition, compromised defecation, and disturbances when engaging in coitus are the most common problems. The hyper-reflexic spastic bladder is believed to be a more significant dysfunction because the high intravesical pressure in the bladder can result in damage to the bladder–ureter countercurrent multiplier system and the kidneys; these major complications can cause death in SCI patients [1].

Restoration of normal micturition function in SCI patients using reconstructive surgery remains a challenge. Conventional treatments included intermittent catheterization, external sphincterotomy, and continent diversion [2, 3], but the results were not reliable. In the 1980s, Brindley et al. introduced electrical stimulators for sacral anterior roots for bladder control in paraplegic patients [4], but this procedure was not widely used because of mechanical malfunction and poor long-term results.

In 1999, Xiao et al. established an artificial skin–central nervous system (CNS)–bladder reflex pathway in a cat model by intradual micro-anastomosis of the left L7 ventral root to S1 ventral root. They demonstrated that this procedure could conduct cutaneous afferent signals to trigger a new micturition reflex arc [5]. The preliminary results of clinical application were satisfactory. This procedure may provide a novel method for controlling bladder function in patients with SCIs and hyper-reflexic bladder, but further investigations are needed.

8.2 Animal Experiments

8.2.1 Introduction

The neurogenic bladder dysfunction caused by SCI has hitherto been difficult to treat. To find a solution to this problem would surely improve the quality of life (QoL) of paraplegic patients and reduce the risk of death. Hou et al. [6–8] based on their experimental study on reconstruction of the Achilles tendon–spinal cord–bladder reflex arc by anastomosis of the S1 and S2 ventral

H. Lin • C. Hou, M.D. (✉)
Department of orthopedic Surgery, Changzheng Hospital, The Second Military Medical University, Fengyang Road 415, Shanghai 200003, People's Republic of China
e-mail: haodonglin@hotmail.com; chunlin_hou@163.com

J. Wang
Department of Orthopedic Surgery, Shanghai Ninth People's Hospital, Shanghai Jiaotong University, Shanghai, People's Republic of China
e-mail: jinwu_wang@sohu.com

G. Zhong
Department of Orthopedic Surgery, Renji Hospital, Shanghai Jiaotong University, Shanghai, People's Republic of China
e-mail: zhonggb2004@yahoo.com.cn

C. Hou (ed.), *Functional Bladder Reconstruction Following Spinal Cord Injury via Neural Approaches*,
DOI 10.1007/978-94-007-7766-8_8, © Springer Science+Business Media Dordrecht 2014

roots, applied this method above the level of the thoracic spine to the spastic bladder caused by SCI and the results were promising. With respect to the frequently occurring atonic bladder due to injury to the conus medullaris caused by thoracolumbar fractures, such artificial reflex arcs were difficult to rebuild.

We wanted to establish an abdominal reflex–spinal cord–central bladder pathway through anastomosis of a healthy ventral root above the injury level and the S2 ventral root with an autologous nerve graft to induce regular micturition. To test this hypothesis, we designed a bladder reflex arc in Sprague–Dawley rats.

8.2.2 Materials and Methods

8.2.2.1 Animals

Twenty male Sprague–Dawley rats (250–300 g) were provided by the Experimental Animal Centre of Fudan University (Shanghai, China). Rats were evenly divided into five groups (A–E groups). In each rat, the right side was the experimental side and the left was the control side.

8.2.2.2 Instrumentation

The SMUPPC Biological Signal Processing system (developed by the Physiology Department of Fudan University and Shanghai Jiaoyi Factory (Shanghai, China)), a silver bipolar hook, a stimulating electrode, a recording electrode, and an intravesical pressure monitor connected to the pressure transducer of the SMUPPC system were used. By using MF software Version 3.00 which is provided from the Physiology Department of Fudan University, all the experimental data would be collected and processed with a high-performance workstation.

8.2.2.3 Establishment of Artificial Bladder Reflex Arc

The rat's lower abdominal reflex center lies within the spinal segments of T12–T13. Among all the relevant nerves, the S2 spinal nerve innervates the bladder with highest potency. Firstly, we anesthetized rats with pentobarbital sodium (concentration: 2 %, 45 mg/kg, i.p.). Then, the rats were fastened on the operating table in supine position. In a full sterile environment, the posterior midline incision ranging from T11 to S4 was operated on the back for exposing the vertebrae. The 13th ribs are the most caudal pair ("floating ribs") and the T13 intercostal nerve is found beneath them. Through tracing to the origin of intercostal nerve, we found the spinal nerve, ventral and dorsal roots of T13. There are four pairs of sacral nerves, among which the S2 spinal nerve is characterized by locating its origin in the L6 spinal nerve near the lumbosacral articulation. In order to expose the dura mater, a right hemilaminectomy was carried out between T11 and S4, as well as the ventral and dorsal roots of T13 and S2. The ventral and dorsal roots of S2 were separated under a surgical microscope. The ventral root of S2 contains only one fasciculus without ganglion and is located on the ventral medial side of the spinal cord. The dorsal root consists of two fasciculi with a spindle-shaped ganglion which is close to the ventral inner side of the intervertebral foramina, and is located on the lateral side of the spinal cord. The S2 ventral root was transected close to the spinal cord, maintaining the integrity of dorsal root and cauda equina. After splitting the T13 nerve root, the T13 dorsal root was shown to be attached the dorsal root ganglion. When plucking the adjacent dorsal root ganglion, the T13 ventral root was exposed. The T13 ventral root was transected against the inner side of the dura mater, keeping T13 dorsal root and dorsal root ganglion intact. By using the surgical microscope, we tracked the T13 ventral root and dissected it along the edge until reaching the spinal cord. Under the surgical microscope which has × 10 magnification, the proximal end of the T13 ventral root was anastomosed to the distal end of the S2 ventral root with autologous nerve grafting of the cocygeal nerve by using an 11–0 atraumatic needle and monofilament nylon sutures. All rats were raised for 8 months before continuing on the following step of the experiment.

8.2.2.4 Electrophysiological Test of Bladder Function

Three groups A-C of rats (each group consists of four rats) were anesthetized by using pentobarbital sodium (concentration: 2 %, 45 mg/kg, i.p.). And then, they were fastened on the operating table in supine position. In a full sterile environment, a middle lower abdominal inverted "Y" incison was carried out under a surgical microscope for revealing the bilateral pelvic splanchnic nerves and its branch named bladder plexus. One side of silver bipolar wire hook electrode, working as recording electrode, was linked to the SMUPPC system. We let the electrode "hook" the bladder plexus, and instilled a drop of mineral oil. The other side of silver bipolar wire needle electrode was stitched into the bladder wall. A ground electrode was placed onto the abdominal muscles. Then, we put the rats in prone position and operated a midline cut along their back as so to expose the extradural scar, eliminate the residual lamina, and disclose the dural capsule. After separating the bilateral ventral and dorsal roots of the T13 and S2 nerves, we put the stimulating electrodes onto the T13 bilateral intercostal nerves (the nerves were divided and the proximal ends stimulated) as well as the left side of the dorsal root of S2. The action potential in the bladder plexus was recorded by the bipolar hook electrode connected to the bladder plexus bilaterally. The single stimulating pulse had 3 mA amplitude and 0.3 ms duration. The compound muscle action potentials (CMAPs) of smooth muscle in the bladder were recorded through the bipolar needle electrode in the bladder wall; the stimulating pulse of 0.3 ms has 3 mA amplitude and 20 Hz frequency in 5 s period. The L5–S4 spinal-cord segment was transected, the spinal nerves left intact, and the action potentials mentioned above recorded again.

8.2.2.5 Intravesical Pressure Test

After carrying out a lower abdominal incision on rats of groups A–C for exposing their bladders, a catheter (diameter, 0.75 mm) was inserted into the bladder through an incision (diameter, 0.7 mm) on the bladder wall with an immobilized stitch. The other end of the catheter was connected to a transducer through using a T-tube, and then measured by the SMUCPC system. Before the test, the bladder and catheter were filled with saline in order to conduct a pressure test by using a mercury manometer. The stimulating electrodes were placed on the bilateral intercostal nerves and the left side of the dorsal root of S2. Intravesical pressure curves were recorded by using a single stimulus of 3 mA in 0.3-ms pulses at 20 Hz for 5 s before and after transection of the L5–S4 spinal-cord segment.

8.2.2.6 Neuropharmacological Tests

After finishing above tests, atropine (0.05 mg/kg, i.m.), which had an effect to block the muscarinic receptor, was injected to the bladder wall of rats in group A; trimethaphan (5 mg/kg, i.m.), which blocks nicotinic receptor-1, was injected to the bladder wall of rats in group B; vecuronium bromide (4 mg/kg, i.m.), which blocks nicotinic receptor-2, was injected to the bladder wall of rats in group C. The intravesical pressure and CMAP were measured with the stimulation of the proximal end of the T13 intercostal nerve on the right side.

8.2.2.7 Preparation of Samples for Microscopy

The tissue from the anastomotic site of T13 and S2 ventral root of group-D rats was excised, embedded in paraffin, sliced lengthwise along the tissue, and stained with hematoxylin and eosin (H&E) so as to observe axon regeneration at the suture site. A median lower abdominal incision was carried out for exposing the bladder and tissues at the bilateral connecting sites of the bladder plexus. The bladder smooth muscle was excised and fastened by glutaraldehyde (2.5 %) and osmic acid (1 %). Fixed tissues were dehydrated in an alcohol gradient, embedded by epoxy resin 618, and cut using an ultramicrotome. They were stained using uranyl acetate and lead citrate for observation under a transmission electron microscope.

8.2.2.8 Retrograde Tracing Using Horseradish Peroxidase (HRP)

In a full sterile environment, a median lower abdominal incision was carried out for exposing the bladders of rats of group E. HRP (5 μL; 30 %; RZ=6.9; Sigma–Aldrich, St Louis, IL, USA) was injected slowly by using a microsyringe into the bladder wall at three points, and the syringe remained there for 10–15 min. Under anesthesia condition lasting 60 min, an incision was made on the chest to expose the heart. A syringe for irrigation was inserted into the left ventricle and ascending aorta. Physiological (0.9 %) saline (50–300 mL) was injected quickly to flush the vessels. After 5 min at room temperature, 0.1 % phosphate-buffered saline mixed with 1 % metaformaldehyde and 1.25 % glutaraldehyde (500 mL) was injected into the vessel: the first 250 mL was injected slowly and the last 250 mL quickly. Irrigation was complete in 30 min when a left hemilaminectomy was carried out to expose the spinal cord. Segments from T11 to L1 and from L4 to S4 were transected and enmeshed in the fixing fluid as mentioned above. In 4–6 h, these segments were dipped in a 0.2 mol/L phosphate buffer and saccharin fluid controlling the temperature at 4 °C. The next day, the tissues were frozen and cut into 40-μm slices. Slides were enmeshed into 0.1 mol/L phosphate buffer for 24 h at 4 °C.

According to the Mesulam tetramethyl benzidine reaction, tissues were mounted on gelatine slides, dehydrated with resin, and then observed for determining the distribution and quantity of HRP (+) cells.

8.2.3 Results

8.2.3.1 Changes in the Action Potential of the Bladder Plexus

By applying the previous electrical stimulus to stimulate the proximal end of the T13 intercostal nerve, the action potential of the bladder plexus was recorded in double sides of the spinal cord transection. We found that the signal waveform was almost the same as record result when the control side of S2 ventral root was excited. However, the action potential was not detected when the same stimulus was applied to the control side of T13 ventral root.

8.2.3.2 Changes in the CMAP of Bladder Smooth Muscle

When stimulating the proximal end of the T13 intercostal nerve, we recorded the CMAP of bladder smooth muscle before and after paraplegia due to the transection of L5–S4 spinal-cord segment. The mean maximum amplitude reached 81 % of that recorded when stimulating the control side of S2 ventral root. The curve was the same as that of the control side. When the identical stimulus was applied on the left side of T13 ventral root, the CMAP of bladder smooth muscle was not induced.

8.2.3.3 Testing of Intravesical Pressure

The intravesical pressure gradually increased and attained maximum value in 1.5–2 s. The mean value of intravesical pressure caused by the artificial reflex arc reached 76 % of that observed on the control side. The curve was the same as that recorded when stimulating on the control side of S2 ventral root. When the identical stimulus was applied on the left side of T13 ventral root, the record result indicated no change in intravesical pressure.

8.2.3.4 Neural Pharmacological Test

After atropine was injected in the bladder wall of group-A rats and trimethaphan was injected in the bladder wall of group-B rats, both the CMAP of bladder smooth muscle and intravesical pressure were not changed when we stimulating the proximal end of the T13 intercostal nerve. After vecuronium bromide was injected in the bladder wall of group-C rats, the CMAP of bladder smooth muscle and intravesical pressure were changed. The results were the same as those recorded before injection. It was indicated that the new reflex arc could be blocked by a muscarinic receptor blocker (atropine) and nicotinic receptor-1 blocker (trimethaphan) but not a nicotinic receptor-2 blocker (vecuronium bromide).

8.2.3.5 Axon regeneration at T13 and S2 Suture Sites

Through an optical microscope, we found out that the neural fibers of T13 ventral-root regenerated into the suture site. That is, axons were arranged in an orderly manner, and neuromas or scar formation were not observed. A new capillary was found at the epineurium, but conspicuous proliferation of connective tissue in the stroma was not observed.

8.2.3.6 Ultramicrostructure of the Ends of Parasympathetic Nerve Fibers in the Bladder Wall

Transmission electron microscopy demonstrated the terminals of parasympathetic nerve fibers without myelin sheath formed by Schwann's cell in the bladder wall between smooth muscle cells (SMCs) and collagen. A differentiated structure similar to the structure of a motor endplate between SMCs and the ends of parasympathetic nerve fibers was not observed. Mitochondria as well as many clear synaptic vesicles and tubules of smooth endoplasmic reticulum were present in the axoplasm of the ends of parasympathetic nerve fibers. Compared with the structure and conformation of the ends of the parasympathetic nerve fibers on the control side, there was no conspicuous difference, except that there were fewer mitochondria and synaptic vesicles on the experimental side.

8.2.3.7 Conformation and Distribution of HRP(+)cells in the Lumbosacral Segment

Optical microscopy revealed HRP(+) cells to be present in bilateral L6–S4 segments. They were distributed mainly in the intermediolateral nucleus zone, and cells appeared to have irregular round or oval shapes. Cells contained abundant deep blue-gray or light blue-gray particles in the cytoplasm, and were prominent around the stained cells.

8.2.3.8 Conformation and Distribution of HRP(+)cells in the T13 Segment

Optical microscopy revealed HRP(+) cells to be present in the ventral horn of the T13 segment and were of different types: maxicells (α motor neuron) were in sight, but medium and tiny cells (α or γ motor neuron) were seldom observed. Cells comprised plentiful blue-gray or light blue-gray particles in the cytoplasm and nucleoli. Positive particles could occasionally be found in the thin prominence around the stained cells. On the control side, HRP(+) cells were found in the ventral horn of T13 segment.

8.2.4 Discussion

SCI can seriously damage the somatic motor and sensory abilities of individuals. Whereas, it has some features to keep the internal organs at or below the damaged level out of control from higher-level nerves (brain and/or spinal cord) and lead to neurogenic dysfunction. The typical dysfunctions in such patients are those of micturition, defecation and copulation, among which the spastic bladder with high tension and high reflection of nervous impulses is the most dangerous symptom. High internal pressure of the bladder has a chance to result in bladder–ureter reflux and damage the kidney in a retrograde fashion, which is the principal cause of death for the late stage of SCI [9, 10]. In the USA, the annual mortality due to SCI is 32 per million population, the annual incidence is 906 per million population, and the injury level is above the conus medullaris in 85 % of cases [11]. Therefore, to discover a solution to this difficult problem, improve the QoL of SCI patients, and to lower the mortality appear to be extremely important tasks.

Frequently used methods for SCI causing a neurogenic bladder include urethral catheterization, pharmacological agents, and urological surgery. These methods are associated with many complication and suboptimal clinical effects. Therefore, reinnervation of the bladder and lower urethral tract has become the focus of research.

In 1976, Brindley et al. [12] utilized electric stimulation therapy of the sacral anterior root, and subsequently the patients recovered satisfactory micturition. In 1994, Xiao et al. [13] implemented an artificial crossover skin–CNS–bladder reflex pathway by intradural microanastomosis of the left L4 ventral root to the L6 ventral root in rats. This reflex pathway innervated the bladder and external urethral sphincter while leaving the L4 dorsal root intact as a trigger for micturition, and gained good results.

We implemented an innovative artificial reflex arc "abdominal reflex–spinal cord–center bladder" through stimulating the lower abdominal skin or the intercostal nerve. Motor impulses were transmitted through the revised efferent branch into the bladder so as to elicit autonomous contractions. According to the results of neural electrophysiological tests and intravesical-pressure tests, this new reflex was verified to be effective. The results of our study were in accordance with others [13–15]. The neural pharmalogical test showed that the abdominal reflex–spinal cord–center bladder artificial reflex arc could be suppressed by muscarinic and nicotinic-1 receptor blockers, but not by nicotinic-2 receptor blockers. These findings suggested that acetylcholine was the neurotransmitter released from the nerve ends of the artificial reflex arc. The nicotinic-1 receptor is used for the ganglion in the bladder wall and the muscarinic receptor is used for the smooth-muscle layer of the bladder wall. Therefore, parasympathetic ganglion neurons as well as their efferent fibers belong to the innovative reflex arc.

In conclusion, an artificial bladder reflex arc could be established through intradural nerve grafting and anastomosis of sacral ventral roots which applied an intact abdominal reflex beyond the height of paraplegia. This arc can realize autonomous micturition for most patients with lumbosacral injury, and provide a new treatment for SCIs caused by thoracolumbar fractures.

Compared to conventional methods for the re-establishment of bladder micturition [16, 17], our method has five main advantages. Firstly, in a certain sense, this abdominal reflex–spinal cord–center bladder artificial reflex arc is close to the normal micturition reflex. Therefore, such an artificial reflex arc can produce synergic bladder contraction and sphincter relaxation, and cause bladder voiding under physiological control. Secondly, choosing the lower abdominal reflex above the lumbosacral SCI level to reconstruct an artificial micturition reflex provides a new method to restore bladder function for patients with atonic bladder caused by SCI due to thoracolumbar fractures. Thirdly, this procedure does not involve electrode implantation or mechanical devices, which can damage nerve roots or cause other complications. Fourthly, this procedure does not require dorsal rhizotomy. Finally, residual sensory function was preserved, which is important for sexual activity and awareness of noxious stimuli. Such reconstruction of an artificial bladder reflex arc needs autonomous nerve grafting. Hence, the quality of the nerve anastomosis and time for axon regeneration into the smooth muscle layer of the bladder wall will influence the final outcome of this newly constituted reflex arc.

We conclude that, using the surviving somatic reflex above the level of the SCI to reconstruct the bladder autonomous reflex arc by intradural micro-anastomosis of ventral roots with a segment of autologous nerve grafting is practical in rats, and may have applications in humans.

8.3 Clinical Applications

8.3.1 Introduction

In our previous experiments [18], we established a new skin–CNS–bladder reflex (abdominal reflex) pathway to restore controlled micturition in the atonic bladder. The new pathway was established in a rat model of SCI by intradural microanastomosis of the right T13 ventral root to the S2 ventral root with autogenous nerve grafting. After the new reflex pathway was reestablished, long-term function of the reflex arc was evaluated through the methods of electrophysiology, detrusor, electromyography, and urodynamics in postoperative 8 months. It is shown that the normal somatic reflex superior to the spinal injury level can be applied to establish a reflex pathway though spinal ventral root anastomosis between the T13 and S2 nerve roots. The somatic motor root reinnervate the bladder with a reconstructed efferent branch. The somatic motor impulses are transmitted to the bladder through the reconstructed efferent branch so as to induce contraction of the bladder detrusor. Relevant results were acquired in a canine model of SCI [19]. Based on the results of these preclinical experiments, we attempted bladder reinnervation in patients with conus medullaris injuries through functional suprasacral nerve transfer.

8.3.2 Indications

Patients who presented with complete conus medullaris injuries with atonia bladder, and who had no contraindications for general anesthesia and surgery with stable neurological injury.

8.3.3 Surgical Technique

The surgical procedure was performed under general anesthesia. A 3-channel Foley catheter was introduced into the bladder, with one channel connected to a drainage bag, the second connected to a pressure transducer leading to an urodynamic unit. A standard laminectomy from L5 to S3 was performed with the patient lying in the prone position. The dura was opened through a paramedian incision, exposing the dorsal and ventral roots of the S2 nerve. The ventral root of S2 was separated from the dorsal root by microdissection and tested by electrostimulation to confirm its location. Electrostimulation of the S2 ventral root (VR) should induce bladder contraction and an increase in intravesical pressure. After its location was confirmed, the S2 ventral root (VR) was transected.

Subsequently, an additional standard laminectomy from T9 to T11 was performed. The T10 spinal segment was identified by the entrance of the T10 roots. After the ventral root of T10 was identified using an operating microscope and found to be functional, it was transected. A sural cutaneous nerve about 30 cm in length was used as the nerve graft. One side of the nerve graft was anastomosed to the proximal stump of the T10 VR using a 10-zero absorbable suture. The nerve graft was then transferred cutaneously to the other incision with a hollow tube and anastomosed to the S2 VR (Fig. 8.1). The wound was closed by suturing in three layers, with an external drain. After surgery, broad-spectrum antibiotics were given for 3 days.

The patients were moved to rehabilitation clinics 2 weeks after surgery. Patient follow-up was scheduled 6, 12, 18, and 24 months after the operation. At these visits, urodynamic studies, including detection of maximum detrusor pressure, maximum flow, and sphincter resistance, were performed and questionnaires were completed. From month 12 onwards, the patients were directed to scratch or gently squeeze the T10 dermatome when the bladder was full to initiate voiding. When there was evidence of significant recovery of bladder function, renal function was re-evaluated.

8.3.4 Case Report

A 43-year-old woman was referred for evaluation and treatment of atonic bladder caused by a conus medullaris injury in July 2000. She had fallen from an elevated height, which resulted in a compression fracture of the L1 vertebra and subsequently complete paraplegia. The patient complained of urinary retention and anuria, which required insertion of an indwelling urinary catheter. As expected, a serious urinary infection had developed, for which antibiotic therapy was instituted. Surgical reconstruction of the bladder reflex pathway was scheduled and performed 4 months after the injury. At this time, the medial sural cutaneous nerve was cut and anastomosed with the T10 and S2 ventral roots (Fig. 8.2).

After the reconstructive procedure, the indwelling urinary catheter was kept in place for 1.5 years and caused multiple urinary system infections. At 2 years after the surgery, the patient experienced recovery of bladder storage and voiding functions. She could void by scratching her T10 dermatome. To date, this patient has been followed for 72 months and the urethral catheter has been removed. She has regained the sensation of bladder fullness, can contract the abdominal muscle, and is capable of complete urination without any difficulty. The total urinary output increased from 500 to 800 ml per day. According to the results of urodynamic analysis, the flow rate of patient's maximal urine was 38 ml/s; urinary output, 596 ml/void; residual urine, 30 ml; vesical pressure, 145 cm H_2O; abdominal pressure, 31 cm H_2O; and detrusor pressure, 114 cm H_2O (Fig. 8.3). The detrusor had acquired the nerve reflex since the external sphincter was denervated. Hence, the recovery of urinary function was mainly contributed to detrusor contraction, instead of bladder compression through the abdominal muscles.

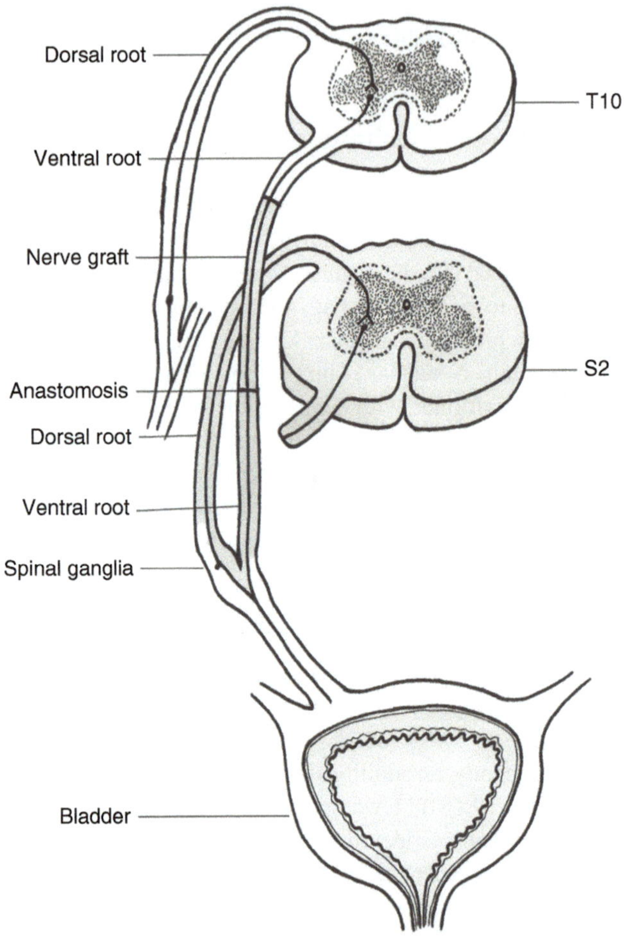

Fig. 8.1 Creation of microsurgical anastomoses between the T10 and S2 ventral roots

8.3.5 Discussion

Based on our preclinical findings in the rodent model [18], we performed spinal ventral root anastomoses in ten paraplegic patients presenting with traumatic lesions of the conus medullaris. Seven of the patients recovered urinary function 2 years after undergoing the surgery of reflex pathway reconstruction [8]. The time course of the reestablished bladder reflex was positively correlated with the time course of axonal regeneration. The conclusions of the post-operative urodynamic study demonstrated that the intravesical pressure rose dramatically, whereas the abdominal pressure had no prominent change. Hence, it concludes that the recovery of urinary function is mainly attributed to detrusor contraction instead of contraction of the bladder which is caused by the abdominal muscles. This summary shows that the effect of therapy was due to the activity in the newly reconstructed bladder pathway. That is, stimulation of the T11 dermatome generated an impulse that was transmitted to the bladder through the regenerated nerve axons, thereby stimulating contraction of the detrusor muscle.

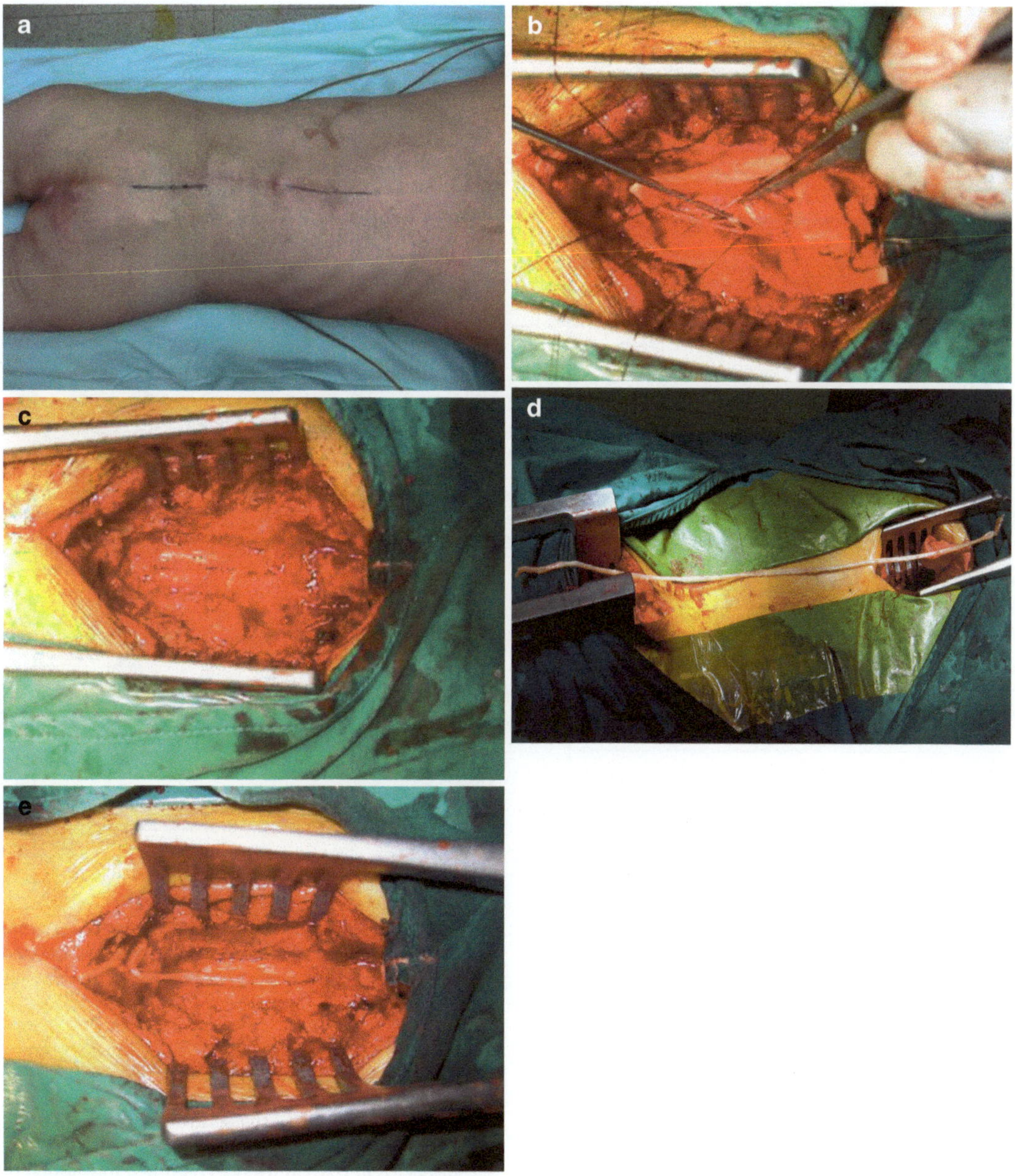

Fig. 8.2 Images of the surgical procedure used for reconstruction of the bladder reflex pathway in a 43-year-old woman. (**a**) Standard laminectomies from L5 to S3 and from T10 to T12 were performed with the patient lying in the prone position. (**b**) After the ventral root of T10 was identified and found to be functional, it was transected. (**c**) The dura was then closed, leaving the T10 ventral root outside. (**d**) A sural cutaneous nerve, about 30 cm in length, was taken for the nerve graft. (**e**) The T10 and S2 ventral roots were then anastomosed through a nerve graft

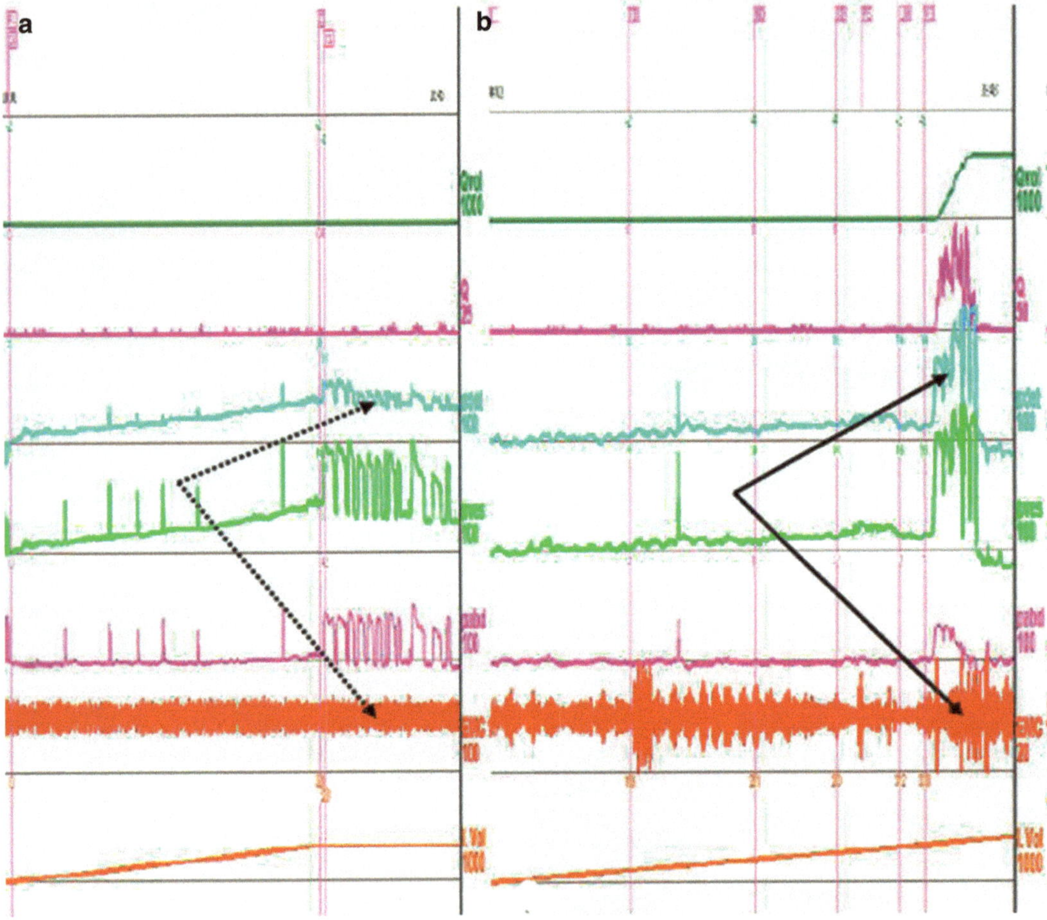

Fig. 8.3 Urine pressure and flow diagram from the urodynamic test. (**a**) Preoperative study showed that the detrusor had no reflection but the external sphincter was denervated. (**b**) Postoperative study showed that the detrusor had regained nerve reflex when the external sphincter was denervated: intravesical pressure increased quickly while abdominal pressure did not increase significantly

References

1. Selzman AA, Hampel N. Urologic complications of spinal cord injury. Urol Clin North Am. 1993;20:453–64.
2. Cass AS, Luxenberg M, Johnson CF, Gleich P. Management of the neurogenic bladder in 413 children. J Urol. 1984;132:521–5.
3. Mitchell ME, Kulb TB, Backes DJ. Intestinocystoplasty in combination with clean intermittent catheterization in the management of vesical dysfunction. J Urol. 1986;136:288–91.
4. Brindley GS, Polkey CE, Rushton DN, Cardozo L. Sacral anterior root stimulators for bladder control in paraplegia: the first 50 cases. J Neurol Neurosurg Psychiatry. 1986;49:1104–14.
5. Xiao CG, de Groat WC, Godec CJ, Dai C, Xiao Q. "Skin-CNS-bladder" reflex pathway for micturition after spinal cord injury and its underlying mechanisms. J Urol. 1999;162:936–42.
6. Hou CL, Zhong HB, Zhang SM, Chen AM, Liu ZD, Liu MX. Establishing an artificial somato-CNS-bladder reflex arc to restore controlled micturition in spinal cord injuries: a preliminary report. Acad J Second Mil Med Univ (Chin). 2000;21:87–9.
7. Lin H, Hou C, Zhen X. Bypassing spinal cord injury: surgical reconstruction of afferent and efferent pathways to the urinary bladder after conus medullaris injury in a rat model. J Reconstr Microsurg. 2008;24:575–81.
8. Lin H, Hou CL, Zhong G, Xie Q, Wang S. Reconstruction of reflex pathways to the atonic bladder after conus medullaris injury: preliminary clinical results. Microsurgery. 2008;28:429–35.
9. Chuang DC, Chang PL, Cheng SY. Root reconstruction for bladder reinnervation: an experimental study in rats. Microsurgery. 1991;12:237–45.

10. Benevento BT, Sipski ML. Neurogenic bladder, neurogenic bowel, and sexual dysfunction in people with spinal cord injury. Phys Ther. 2002;82:601–12.
11. Frankel HL, Coll JR, Charlifue SW, Whiteneck GG, Gardner BP, Jamous MA, et al. Long-term survival in spinal cord injury: a fifty year investigation. Spinal Cord. 1999;36:266–74.
12. Brindley GS. The first 500 patients with sacral anterior root stimulator implants: general description. Paraplegia. 1994;32:795–805.
13. Xiao CG, Godec CJ. A possible new reflex pathway for micturition after spinal cord injury. Paraplegia. 1994;32:300–7.
14. Sievert KD, Xiao CG, Hennenlotter J, Seibold J, Merseburger AS, Kaminskie J, et al. Voluntary micturition after intradural nerve anastomosis. Urology A. 2005;44:756–61.
15. Hou CL, Zhong HB, Zhang SM, Liu MX. Experimental study on establishment of artificial bladder reflex arc after spinal cord injury. Chin J Reparative Reconstr Surg (Chin). 2000;14:10–3.
16. Zheng XY, Hou CL, Zhong HB, Xu RS, Chen AM, Xu Z. Reconstructed bladder innervation below the level of spinal cord injury: the kneetendon to bladder artificial reflex arc. J Spinal Cord Med. 2009;32:79–85.
17. Wang HZ, Li SR, Wen C, Xiao CG, Su BY. Morphological changes of cholinergic nerve fibers in the urinary bladder after establishment of artificial somatic autonomic reflex arc in rats. Neurosci Bull. 2007;23:277–81.
18. Wang JW, Hou CL, Lu N. Long-term functional study on the artificial bladder reflex arc established by abdominal reflex after spinal cord injury. Chin J Microsurg. 2002;25:280–4.
19. Zhong GB, Hou CL, Wang SB. Experminent study on the artificial bladder reflex arc established in therapy of flaccid bladder after spinal cord injury. Chin J Repar Reconstr Surg. 2006;20:812–5.

Reconstruction of Afferent and Efferent Nerve Pathways of the Atonic Bladder

9

Haodong Lin and Chunlin Hou

9.1 Background

Restoration of normal micturition in SCI patients is a major goal in neuro-urology. The ideal therapy would be to achieve repair of the injured spinal cord, but this is not feasible at the present time. Thus, other procedures have been deployed, including intermittent catheterization, electrical stimulation of the sacral ventral roots, continence diversion and external sphincterotomy [1–3]. Unfortunately, these approaches have hardly led to satisfactory outcomes in the majority of SCI cases. Recently, Xiao et al. reported a series of experimental and clinical studies attempting to establish a "skin-central nervous system (CNS)-bladder" reflex pathway to trigger bladder contraction [4–6]. However, this technique was useful only in the cases of spastic bladder incurred by supraconal injuries. Clinically, fracture-dislocation of the spinal column was most commonly witnessed in the thoracolumbar region (T12/L1), This is more likely to result in damage to the spinal cone and bladder center, triggering hypotonia and decreased bladder contractile strength. This study was performed in an attempt to demonstrate that restoration of autonomic micturition in the atonic bladder by re-establishment of afferent and efferent nerve pathways is achievable.

H. Lin • C. Hou, M.D. (✉)
Department of Orthopedic Surgery, Changzheng Hospital, The Second Military Medical University, Fengyang Road 415, Shanghai 200003, People's Republic of China
e-mail: haodonglin@hotmail.com; chunlin_hou@163.com

9.2 Animal Experiments

9.2.1 Introduction

The bypass of a spinal cord lesion using a suprasacral nerve transfer has been demonstrated in both animal and human studies [7–11]. Although not all results have been positive, some have been encouraging [12]. After a careful review of previous work in which bladder reinnervation has been attempted, as well as a consideration of the neuroanatomic characteristics of the spinal nerves, we aimed to reconstruct both the afferent and efferent nerve pathways of the atonic bladder using nerve transfer techniques.

9.2.2 Material and Methods

9.2.2.1 Animal Preparation and Surgical Procedure

A total of 30 male Sprague–Dawley rats were used in this experiment. The animals were housed three per cage with a 12 h light/dark cycle, and were fed rat chow and water *ad libitum*. A left-sided lateral hemilaminectomy was performed under sterile technique to expose the dorsal root ganglia from L4 to S3. The left-side L5 ventral root (VR) was cut distally before it emerged from the dura mater, while the S2 VR was cut more

C. Hou (ed.), *Functional Bladder Reconstruction Following Spinal Cord Injury via Neural Approaches*, DOI 10.1007/978-94-007-7766-8_9, © Springer Science+Business Media Dordrecht 2014

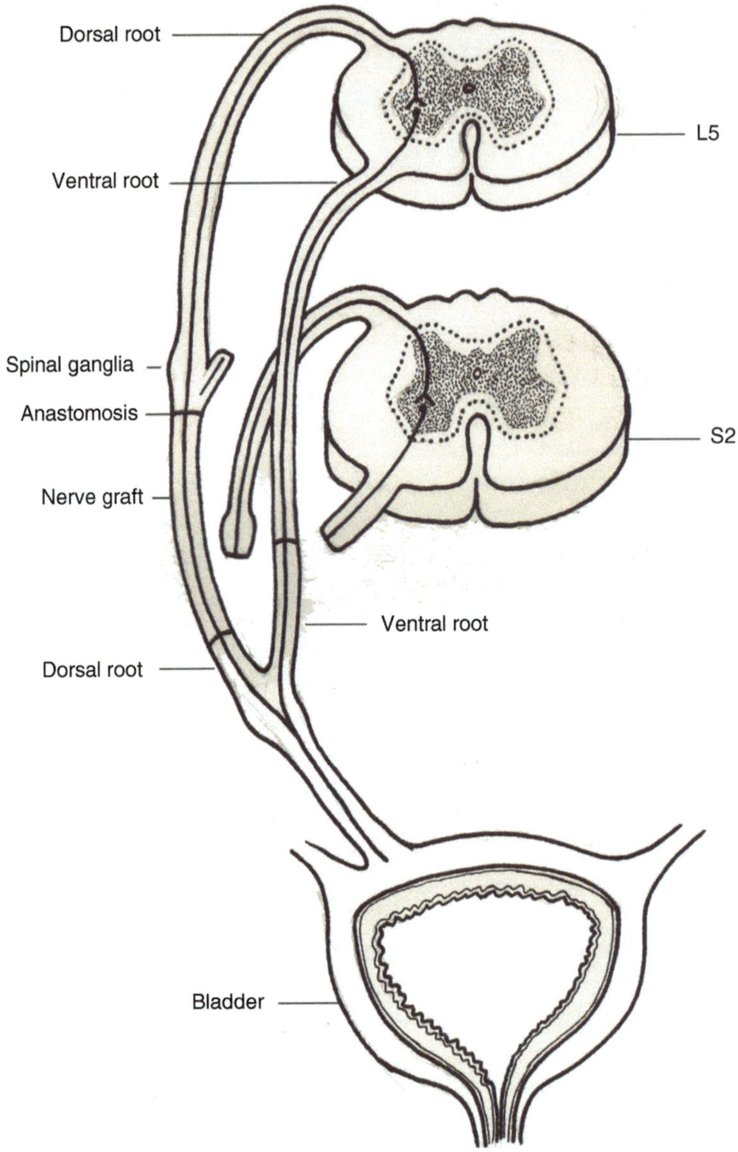

Fig. 9.1 Illustration of relevant neuro-urological relationships for the pure motor-to-motor and pure sensory-to-sensory nerve reconstruction procedure

proximally to enable the direct coaptation of the nerve root ends using two sutures with 12–0 absorbable material. The postganglionic L5 spinal nerve was then cut as distally as possible. The S2 spinal nerve was identified and cut at a point that was immediately distal to the ganglion but proximal to the confluence of the dorsal and ventral roots. A nerve graft (approx. 1.0 cm in length, harvested from the coccygeal nerve) was used to coapt the distance between the proximal stump of the L5 spinal nerve and the distal stump of the S2 sensory root. The roots on the right side was preserved intact and served as an internal control (Fig. 9.1). The dura was then closed with 11–0 nylon sutures, and the wound was closed in layers. Surviving rats were evaluated after an 8-month period.

9.2.2.2 Measurement of Internal Bladder Pressure

After a vertical lower abdominal incision was performed and the bladder was exposed, a 1.0-mm incision was made in the bladder wall and a 0.75-mm diameter catheter was inserted. Then, a pressure monitoring probe was put into the bladder to measure the internal vesicular pressure. The pressure monitoring probe was linked to a pressure transducer that sent the pressure signal to a SMUP-E biological signal processing system (BSPS; Department of Physiology and Pathophysiology, School of Medicine, Fudan University, Shanghai, China). The stimulating electrodes were put on the S2 dorsal roots (DRs) bilaterally. The intravesicular pressure curves were recorded before and after the L6-S4 spinal cord segments were transected via the SMUC-E biological signal processing system. The stimulus intensity was 3 mA, with a pulse width of 0. 3 ms, a frequency of 20 Hz and a duration of 5 s.

9.2.2.3 Electrophysiological Studies

The bladder and bladder nerve plexus of each rat were exposed through a vertical incision on the lower abdominal wall. Then, a silver bipolar electrode was hooked onto the bladder nerve plexus, and a further silver bipolar electrode was sutured onto the bladder wall using 6–0 nylon suture material. After that, the rat was placed in the prone position. The VR and DR anastomoses of the L5 and the S2 nerves were carefully dissected from surrounding scar tissue via the original incision and exposed with the assistance of an operating microscope. The stimulating electrodes were placed on the S2 DRs bilaterally. Bladder plexus action potentials were recorded via the bipolar hook electrode that was connected to the bladder plexus. An extra ground electrode was fixed to the abdominal musculature of the rat. The resulting action potentials in the bladder nerve plexus were examined using a stimulation intensity of 3 mA and a stimulation period of 0.3 ms. Compound muscle action potentials of bladder smooth muscle were recorded with a stimulation intensity of 3 mA, a pulse width of 0. 3 ms, a frequency of 20 Hz and a duration of 5 s. The L6-S4 spinal cord segment was then surgically transected, leaving the spinal nerves intact, and repeated recordings were made of the action potentials described above.

All signals were digitalized and analyzed using MFLab software Version 3.01 (Department of Physiology and Pathophysiology, School of Medicine, Fudan University).

9.2.2.4 Histological Examination

All rats were sacrificed after the electrophysiological studies were completed. Specimens that included the proximal and distal stumps of nerve repair sites were harvested and prepared for histological examination. The nerves were fixed overnight in a solution of phosphate-buffered 2.5 % glutaraldehyde, after which samples were serially dehydrated with gradient alcohols and embedded in Spurr's resin. Two 2-mm long segments were obtained 1 mm proximal and 1 mm distal from the nerve graft repair site and were sectioned. Transverse sections that were 1-μM thick were obtained and stained with 1 % toluidine blue. One section from every three cuts was stained. A total of five sections from the proximal segment and five sections from the distal segment of each nerve were analyzed for the total number of myelinated axons. To determine the total number of axons, each histological section was digitally photographed (Panasonic WV-CP410, Panasonic, Japan) together with an optical microscope (Leica DWLB2, Strasse, Austria). Leica FW4000 Image Analysis Software (Leica Microsystem) was employed to conduct the neuronal morphometric analysis, and the axon counts in all fields were then averaged. An index representing the percentage of axons that crossed the repair site was calculated according to the following formula:

$$\frac{\text{Average number of myelinated axons in the distal segment}}{\text{Average number of myelinated axons in the proximal segment}} \times 100$$

9.2.2.5 Statistical Analysis

Results were interpreted as the mean ± standard deviation. SPSS 13.0 software (SPSS Inc., Chicago, IL) was adopted for the statistical analysis. The analysis of variance (ANOVA) was utilised to compare differences between groups, and results were considered as significant value when p-values were <0.05.

9.2.3 Results

Sixteen animals were excluded from the analysis as they had to be sacrificed during the follow-up period. These animals had developed chronic health problems, which were evidently caused by complications from the procedure such as infected pressure sores or severe and recurrent urinary tract infections. The results presented below are based on the remaining 14 rats that remained well generally during the follow-up period.

9.2.3.1 Action Potentials of the Vesicular Nerve Plexus

Bladder plexus action potentials were recorded both before and after spinal cord transection using the same electrical stimulus to excite the distal end of the left L5-S2 dorsal root anastomosis distal to the nerve graft. The amplitude of the evoked potentials was 0.11±0.02 mV prior to transection and 0.12±0.04 mV after transection, which was non-significant. The shape of the curve recorded on the left side was similar to that on the control side when the right-sided S2 DR was excited (Fig. 9.2). The amplitude of the evoked potentials on the control side was 0.15±0.03 mV, which was larger than those of the left side ($p<0.01$). Action potentials were not detected when the same stimulus was applied to the right-sided S2 DR after spinal cord transection.

9.2.3.2 Compound Muscle Action Potentials in Bladder Smooth Muscle

Compound muscle action potentials of bladder smooth muscle were recorded both before and after the onset of paraplegia triggered by L6-S4 transection by stimulating the distal end of the L5-S2 dorsal root anastomosis on the experimental side. The mean maximum amplitudes before and after paraplegia were 0.12±0.03 mV and 0.12±0.02 mV, respectively, which was not a significant difference. The shape of the curve was similar to that of the control side (Fig. 9.3). The amplitude of the right-sided bladder smooth muscle compound muscle action potential was 0.17±0.03 mV, which was significantly larger than those on the left side ($p<0.01$). When the

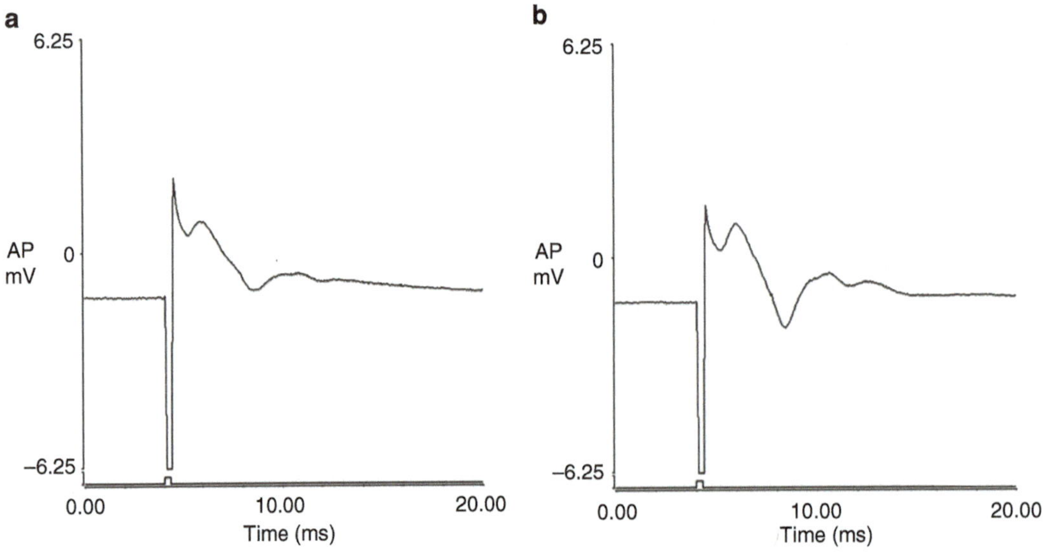

Fig. 9.2 Bladder plexus action potentials were recorded both before and after transection of the spinal cord by using the same electrical stimulus to excite the distal end of the left-sided L5-S2 dorsal root anastomosis. The morphology of the tracing recorded on the left side was similar to that from the control side. The amplitude of the evoked potentials on the control side was larger than those of the left side. (**a**) control side; (**b**) left side

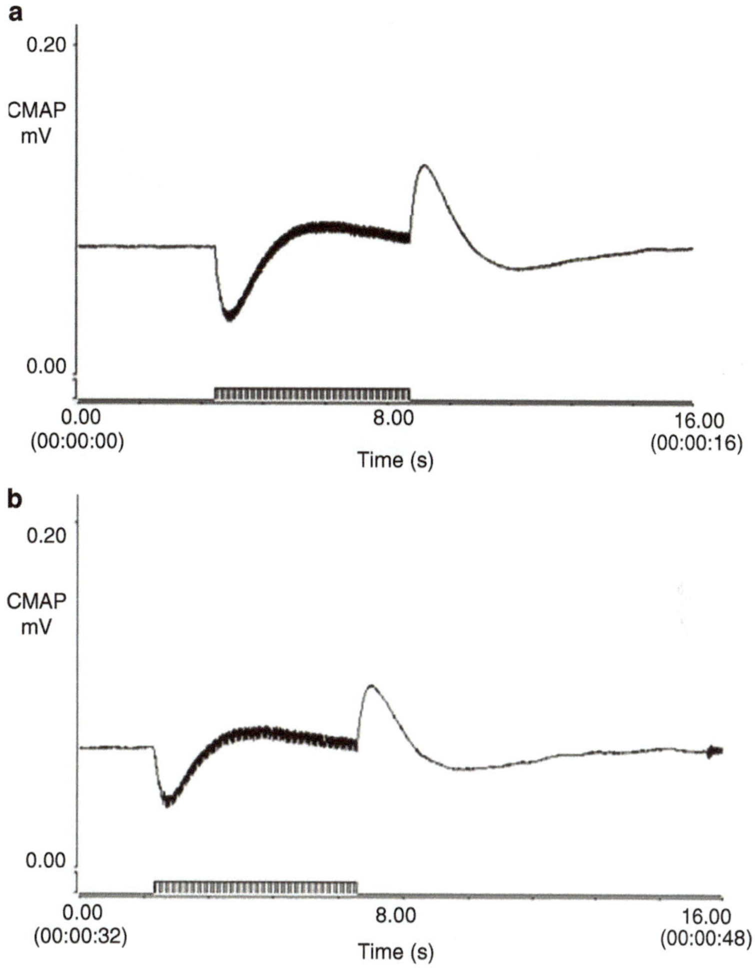

Fig. 9.3 Compound muscle action potentials of the bladder smooth muscle were recordable both before and after the onset of paraplegia caused by L6-S4 transection by stimulating the distal end of the L5-S2 dorsal root anastomosis on the experimental side. The morphology of the tracing was similar to that of the control side. The amplitude of the compound muscle action potentials of the right-sided bladder smooth muscle was significantly larger than that of the left side. (**a**) control side; (**b**) left side

same stimulus was applied to the right-sided S2 DR after paraplegia was induced, a compound muscle action potential could not be elicited from the bladder smooth muscle.

9.2.3.3 Intravesicular Pressure

When the L5-S2 dorsal root anastomosis was stimulated on the experimental side, intravesicular pressure increased both before and after the induction of paraplegia. The recorded intravesicular pressure curve increased steadily and slowly, reaching its peak within 1.5–2.0 s. The mean intravesicular pressures elicited by the artificial reflex arc before and after the induction of paraplegia were 6.57 ± 1.38 cmH_2O and 6.24 ± 1.98 cmH_2O, respectively, which was not significant. The shape of the curve was similar to that recorded on the control side when the S2 DR was stimulated before spinal cord transection (Fig. 9.4). The mean intravesical pressure was 10.88 ± 1.92 cmH_2O on the control side, which was significantly higher than that of the left side ($p < 0.01$). When the same stimulus was used on the right-sided S2 DR after spinal cord transection, no change was recorded in intravesicular pressure.

9.2.3.4 Histological Examination

All specimens of the distal portions of the nerve grafts were observed to have prominent axonal sprouting on histological examination. The distal-to-proximal axon regeneration ratios of the VRs and DRs were 69.85 ± 6.27 % and 55.23 ± 5.37 %, respectively.

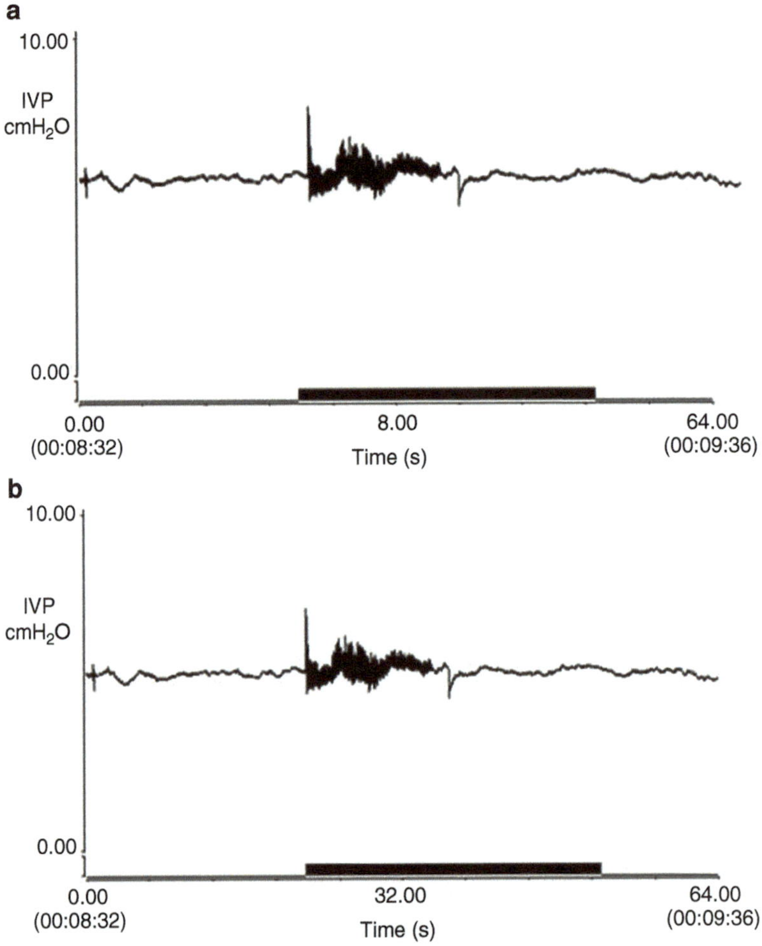

Fig. 9.4 Intravesicular pressure increased by stimulating the L5-S2 dorsal root anastomosis on the experimental side both before and after paraplegia was induced. The slope of the curve was similar to that recorded on the control side. The mean intravesicular pressure of the control side was higher than that of the left side. (**a**) control side; (**b**) left side

9.2.4 Discussion

We used a rat model based upon a previous report using a protocol that was designed to enable an evaluation of the functionality of the reconstructed reflex pathway [6]. Our methodology was similar to that described by Chuang et al. [13] but with certain modifications. For example, it is extremely difficult to keep a paraplegic animal alive for a long duration, the nerve pathway was reconstructed prior to spinal cord transection, which could explain the lower mortality rate observed in our series.

Theoretically, pure motor-to-motor and sensory-to-sensory nerve reconstruction is the best way to reinnervate the bladder. Our studies illustrated that both afferent and efferent nerve pathways can be established by anastomoses of the spinal VRs and DRs between the L5 and S2 nerve roots. Then, the somatic nerve root can reinnervate the bladder through the reconstructed afferent and efferent branches. The application of electrical stimuli to the left L5-S2 DR anastomosis led to motor evoked potential responses. These were recorded in the bladder plexus before and after transection of the rat spinal cord between the L6 and S4 segments. The shapes of the resultant evoked potential curves were similar to those of the control sides, and there was no evidence of interference by non-native nerve root fibers. Urodynamic results demonstrated that the bladder pressure increased after paraplegia when stimulation was applied to the let L5-S2 DR anastomosis, which was approximately 56 % that recorded from the normal controls. The 14 rats all displayed histological evidence of nerve regeneration. In combination, these results demonstrated that new nerve pathways had been established successfully.

9.3 Clinical Applications

9.3.1 Introduction

In our previous experiments in a rat model, we confirmed that it is possible to achieve suprasacral nerve motor-to-motor and sensory-to-sensory nerve transfers after conus medullaris injuries to reconstruct the bladder autonomic reflex arc [13]. Specifically, these were achieved by simultaneous microanastomoses of both VRs and DRs between L5 and S2. Based on these preclinical studies, we then attempted to re-innervate the atonic bladder of patients who suffered previous conus medullaris injuries with the use of suprasacral nerve motor-to-motor and sensory-to-sensory nerve transfers.

9.3.2 Indications

Patients were candidates if they had presented with bladder atonia after a conus medullaris injury.

9.3.3 Surgical Technique

Under general anesthesia, a standard laminectomy was performed from L5 to S3 with the patient lying prone. The dura was opened via a paramedian incision, which exposed the dorsal and ventral roots of the S1-S4 nerves. The S1 nerve root was located using the L5/S1 intervertebral space as a marker, after which the S2-4 nerve roots could be located in descending order. The VRs and DRs at the dural incision were identified according to their anatomical characteristics, and those of S1 and S2 were identified by further microdissection. An electrical stimulator was used to stimulate the S1 VR and lower limb muscle contractions were observed to verify the root identification was correct. If the function of the S1 root was normal, the unilateral S1 and S2 VRs and DRs were transected microsurgically and anastomosed with 9–0 suture material (Fig. 9.5). The wound was then sutured in three layers and an external drain was left in situ.

9.3.4 Case Report

A 40-year-old male was referred for the treatment of an atonic bladder that had been caused by a conus medullaris injury in July 2008 when he fell from an elevated height. This had resulted in a compression fracture of the L1 vertebra, and he had developed bladder atonia with preserved motor functions of the lower extremities. Physical examination revealed that his leg strength was slightly weak but he could walk steadily. Both spontaneous voiding and his sense of bladder fullness were absent, and cystometrography revealed that the patient had an areflexic bladder. Surgical reconstruction of the bladder reflex pathway was performed 8 months after injury. The bladder reflex pathway was established by the intradural microanastomosis of the S1 VR to the right-sided S2 VR to restore pure motor-to-motor reinnervation coupled with S1 DR to the S2 DR to achieve pure sensory-to-sensory reinnervation.

The patient regained bladder sensory function 10 months postoperatively, which meant he had the ability to sense a full bladder and perceive the desire to void. He could void voluntarily, and a comparison of the pre- and postoperative urodynamic study results demonstrated significant improvements in bladder function by the tenth postoperative month. He eventually experienced full recovery of his bladder storage and voiding functions, with no nocturnal urinary incontinence.

9.3.5 Discussion

The results of the current series of patients suggested that a new urination center above the level

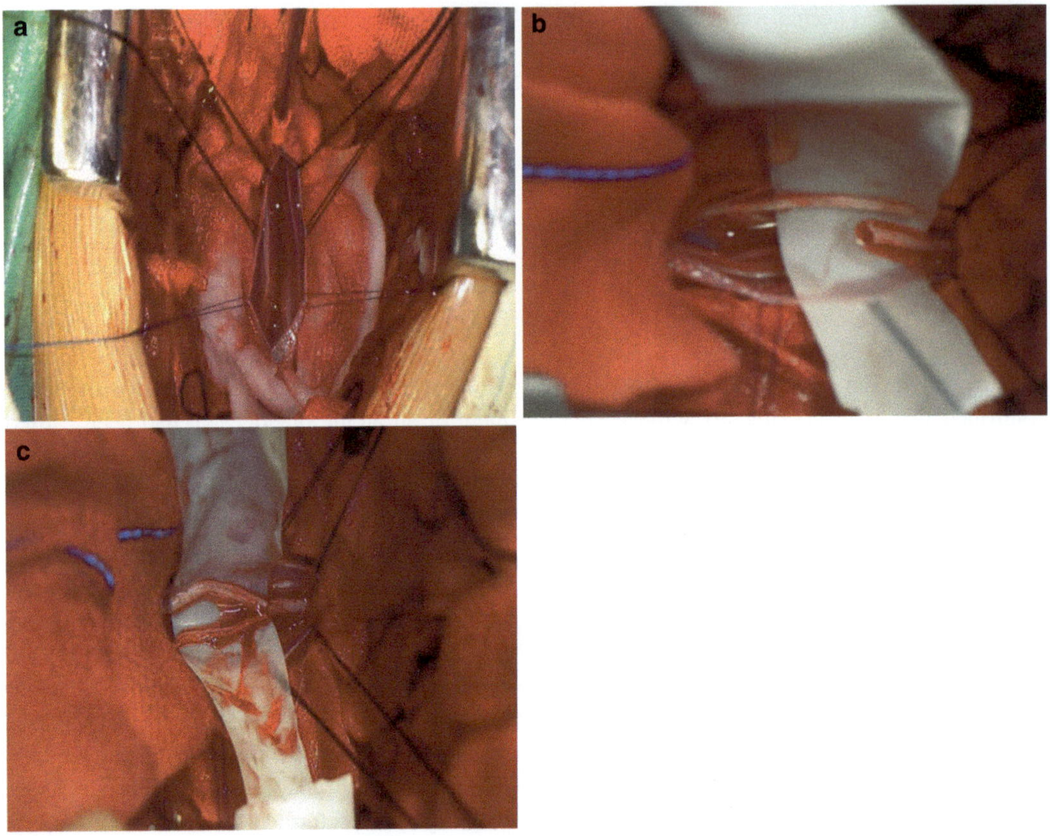

Fig. 9.5 Surgical procedure to reconstruct bladder reflex pathway in a 40-year-old man. (**a**) Standard laminectomies were performed from L5 to S3 with the patient in the prone position. The dura was opened through a paramedian incision, which exposed the dorsal and ventral roots of the S1 and S2 nerves. (**b**) The ventral and dorsal roots of S1 and S2 were then identified and separated by microdissection. (**c**) S1 ventral and dorsal roots were anastomosed to S2 ventral and dorsal roots, respectively

at which paraplegia occurred may have been established after the reconstruction of the afferent and efferent nerve pathways of the atonic bladder. The spastic bladder incurred by spinal injury above the level of the conus medullaris can utilise the latent residual tendon reflex below the plane of paraplegia to rebuild a new "tendon-spinal cord-bladder" reflex if a spinal nerve anastomosis if formed [14]. As the tendon reflex below the plane of paralysis is deployed, it can hardly connect with the cerebral center. We therefore suggest that a "trigger point" may have been formed in this procedure, which indicates bladder fullness and the urge for urination if the trigger point is stimulated. However, it seems to be possible to establish a new pathway for the atonic bladder above the level of spinal cord transection that keeps a cerebral connection. This enables the patient to have more volitional control of urination without the need for trigger point stimulation. In this case, the impulse produced is transmitted to the cerebral micturition center when the bladder is filled.

References

1. Brindley GS, Polkey CE, Rushton DN. Sacral anterior root stimulator for bladder control in paraplegia: the first 50 cases. J Neurol Neurosurg Psych. 1986;49:1004–14.
2. Tanagho EA, Schmidt RA. Electrical stimulation in the clinical management of the neurogenic bladder. J Urol. 1988;140:1331–9.
3. Mitchell ME, Kulb TB, Backes DJ. Intestinocystoplasty in combination with clean intermittent catheterization in the management of vesical dysfunction. J Urol. 1986;136:288–91.

4. Xiao CG, Godec CJ. A possible new reflex pathway for micturition after spinal cord injury. Paraplegia. 1994;32:300–7.
5. Xiao CG, de Groat WC, Godec CJ, Dai C, Xiao Q. "Skin-CNS-bladder" reflex pathway for micturition after spinal cord injury and its underlying mechanisms. J Urol. 1999;162:936–42.
6. Xiao CG, Du MX, Dai C, Li B, Nitti VW, de Groat WC. An artificial somatic-central nervous system-autonomic reflex pathway for controllable micturition after spinal cord injury: preliminary results in 15 patients. J Urol. 2003;170:1237–41.
7. Vorstman B, Schlossberg S, Landy H, Kass L. Nerve crossover technique for urinary bladder rennervation: animal and human cadaver studies. J Urol. 1987;137: 1043–7.
8. Chuang DC, Chang PL, Cheng SY. Root reconstruction for bladder reinnervation: an experimental study in rats. Microsurgery. 1991;12:237–45.
9. Vorstman B, Schlossberg S, Kass L, Devine CJ. Urinary bladder reinnervation. J Urol. 1986;136:964–9.
10. Carlsson CA, Sundin T. Reconstruction of efferent pathway to the urinary bladder in a paraplegic child. Rev Surg. 1967;24:73–6.
11. Sundin T, Carlsson CA. Reconstruction of severed dorsal roots innervating the urinary bladder. An experimental study in cats. II. Regeneration studies. Scand J Urol Nephrol. 1972;6:185–96.
12. Vorstman B, Schlossberg S, Kass L. Investigations on urinary bladder reinnervation: historical perspective and review. Urology. 1987;30:89–96.
13. Lin H, Hou C, Zhong G, et al. Reconstruction of reflex pathways to the atonic bladder after conus medullaris injury: preliminary clinical results. Microsurgery. 2008;28:429–35.
14. Lin H, Hou C, Zhen X, Xu Z. Clinical study of reconstructed bladder innervation below the level of spinal cord injury to produce urination by Achilles tendon-to-bladder reflex contractions. J Neurosurg Spine. 2009;10:452–7.

Transfer of Normal Lumbosacral Nerve Roots to Reinnervate Atonic Bladder

10

Haodong Lin, Zhen Xu, and Chunlin Hou

10.1 Background

With the rapid development of the building industry and of transportation in recent years, the number of cases of spinal cord injury (SCI) is rising each year. Urine retention and bladder dysfunction arising from SCI seriously affect the quality of life and survival of these patients. The total incidence of urinary tract infection in paraplegia patients (1,003 patients) after the Tangshan earthquake (1976) in China was 41.58 %, 49–66 % of whom succumbed to uremia. In the United States alone, more than half a million patients suffer from neurogenic bladder after SCI, and an additional 50,000 people sustain an SCI each year [1]. Other studies report that recurrent urinary tract infections, hypertension, and renal failure, which is one of the major complications, causes death in late-stage SCI patients [2].

Restoration of normal micturition in SCI patients is very difficult as repair of the injured spinal cord is not feasible at the present time. Other approaches have been used, such as intermittent catheterization, external sphincterotomy and continence diversion [3], which can improve urine retention, but the bladder detrusor remain dysfunctional. Since it came into use, the Brindley Sacral Anterior Root Stimulator has allowed paraplegic patients with spastic bladder to regain urinary continence and a satisfactory quality of life in most cases [4]. Recently, Xiao et al. reported a series of experimental and clinical studies attempting to establish a "skin-central nervous system (CNS)-bladder" reflex pathway to trigger bladder contraction [5–7], and we also reported a study to produce urination by Achilles tendon-to-bladder reflex contractions [8, 9]. Although they have proven effective in some instances, these techniques have been used only in cases of spastic bladder. However, fracture-dislocation of the spinal column was most commonly seen in the thoracolumbar region (T12/L1), which can destroy the conus medullaris, resulting in atonic bladder. Consequently, in our previous studies we succeeded in establishing a skin-CNS-bladder reflex (abdominal reflex) pathway for restoring controllable micturition in the atonic bladder [10–12]. Another study demonstrated that reestablishment of afferent and efferent nerve pathways to restore autonomic micturition in the atonic bladder is achievable in a rat model [13]. Using the normal abdominal reflex above the level of injury to reconstruct bladder reflex arc in atonic bladder ascribe to cone injury has been performed clinically [12]; however, this requires a sural nerve transplant, which is nearly 30 cm long, to bridge the T_{11} and S_2 nerve root. The disadvantages of this procedure include difficulty of nerve transplantation, severe surgical trauma and a long period of recovery due to the length of the nerve

H. Lin • Z. Xu • C. Hou, M.D. (✉)
Department of Orthopedic Surgery, Changzheng Hospital, The Second Military Medical University, Fengyang Road 415, Shanghai 200003, People's Republic of China
e-mail: haodonglin@hotmail.com; hawk3412@hotmail.com; chunlin_hou@163.com

C. Hou (ed.), *Functional Bladder Reconstruction Following Spinal Cord Injury via Neural Approaches*,
DOI 10.1007/978-94-007-7766-8_10, © Springer Science+Business Media Dordrecht 2014

transplanted. Furthermore, if the time over which the atonic bladder remains denervated is too long, the detrusor muscle and its neuromuscular junction may degenerate, and it will then probably be of no use for the operation. Hence, a method to more quickly recover the detrusor should be used for atonic bladder.

10.2 Animal Experiment

10.2.1 Introduction

We find that a percentage of patients with atonic bladder due to conus medullaris injury (such as L_1 fracture) present with clinical bladder dysfunction while most or even all of the motor function of the lower extremities remains. For such patients, we wondered whether it would be possible to use the nerve root dominating the normal motor function of the lower extremities as the motor nerve root, i.e., anastomosing the central end of the S_1 anterior nerve root of the unaffected spinal cord segment with the peripheral end of the S_2 and S_3 anterior nerve root that dominates the bladder, to reconstruct an artificial tendon-to-bladder reflex arc for the recovery of bladder function. However, this raises the question of whether severing the S_1 nerve root would result in new functional injury to the patient's lower extremities. This preliminary study was performed to answer this question.

10.2.2 Material and Method

10.2.2.1 Animal Preparation and Surgical Procedure

A total of 40 adult male Sprague–Dawley (SD) rats, weighing 200–220 g, were randomized into four groups of 10 rats each. The animals were housed three per cage under a 12-h light/dark cycle, and were fed rat chow and water ad libitum. All the procedures were approved by the University Animal Care and Use Committee. All the animals were treated in accordance with the Animal Care Guidelines of the National Bureau of Health. Considering that the nerve roots that dominate the sacral plexus (involving L_3–L_6 and predominantly the L4, L5 and L6 nerve roots) [13] may vary between individual SD rats, the right L4, L5 and L6 were severed, respectively, in the three experimental groups designated as groups B, C and D. The normal group was designated as group A. Pre- and postoperative electromyograms (EMG), appearance and gait were observed and compared. Instruments used included an electromyograph and silver needle electrodes.

To prepare for surgery, the animals were anesthetized by intraperitoneal injection of 1 % sodium thiopental (5 mg/kg body weight; Shanghai XinYa Pharmaceutical Co., Shanghai, China). The animals were fixed in a prone position on a plate. Under sterile conditions, a longitudinal incision was made on the right lateral thigh to incise the skin and subcutaneous tissue and isolate the deep fascia, to locate the sciatic nerve along the intermuscular space. The stimulation electrodes were hooked to the sciatic nerve and the recording electrodes were located in the triceps surae muscle, anterior tibialis muscle and biceps femoris, to record EMGs. The L6 spina was located based on the sacral iliac joint, above which were the L4 and L5 spina. A longitudinal incision was made using L4, L5 and L6 as the center. The skin, subcutaneous tissue and deep fascia were incised to expose the spina and the articular processes. Along the right articular process, the transverse process was isolated downward and partially excised to expose the nerve root (Figs. 10.1 and 10.2). The right L4, L5 and L6 nerve roots, respectively, were then severed in groups B, C and D. The left side was kept intact. Finally, the wound was closed in layers. The same surgery was performed in group A, but without severing of the nerve root. Then, at 4, 8 and 12 weeks post-operation, the sciatic nerve was again exposed and stimulated by the same process.

10.2.2.2 Electrophysiological Studies

Following induction of anesthesia, the rat was placed in the prone position. The stimulation electrodes were hooked to the sciatic nerve and the recording electrodes were located in the triceps surae, anterior tibialis and biceps femoris, respectively, to record EMGs. One recording

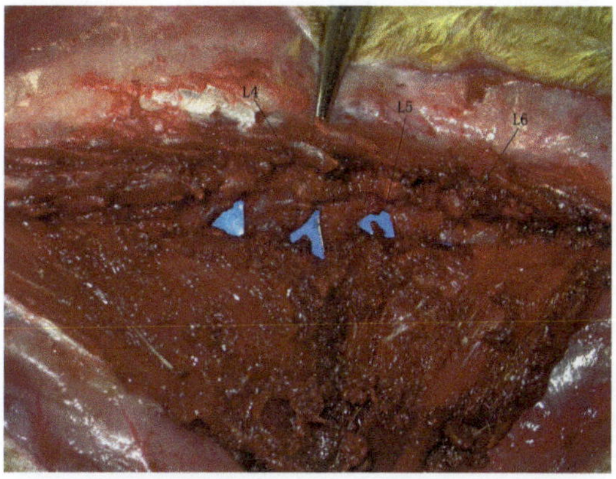

Fig. 10.1 Expose the L4, L5 and L6 nerve root, each was indicated by a line

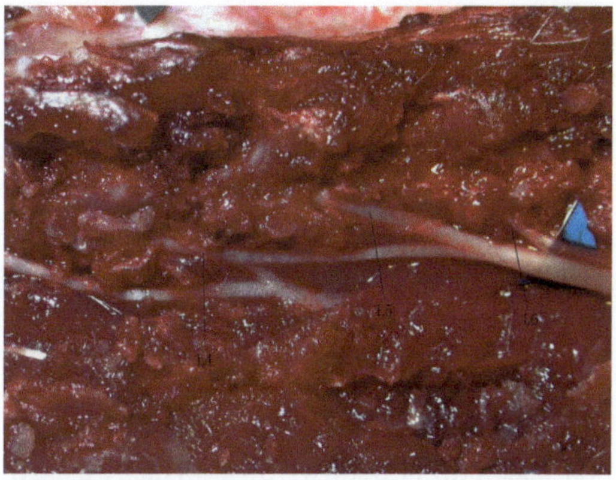

Fig. 10.2 Completely explosion, each nerve root was indicated by a line

electrode was buried in the muscle belly, another in the muscle tendon, and the remaining electrode was fixed to the gluteus maximus as an extra ground electrode. The compound muscle action potentials of the triceps surae, anterior tibialis and biceps femoris were recorded with a stimulating intensity of 1 mA, a pulse width of 0.2 ms, a frequency of 1 Hz, and duration of 10 ms. Before one of the right L4, L5 and L6 nerve roots was severed in each experimental group, the sciatic nerve was stimulated and the compound muscle action potentials of the triceps surae, anterior tibialis and biceps femoris were recorded on the computer. After the nerve root was severed, the same operation was performed and the compound muscle action potentials were again recorded. Then, at 4, 8, and 12 weeks post-operation, the sciatic nerve was stimulated and the compound muscle action potentials were recorded by the same process.

All signals were imported to a computer and analyzed using MFLab software version 3.01 (Department of Physiology and Pathophysiology, School of Medicine, Fudan University, Shanghai, China).

10.2.2.3 Sciatic Functional Index (SFI)

Pre- and post-operation (4-, 8-, and 12-weeks), walking tracks were obtained before induction of anesthesia. Animals were tested in a confined walkway 42-cm in length and 8.2-cm in width, with a dark shelter at the end. A sheet of white paper was placed on the floor of the walking corridor. After the hind paws of the rats were pressed down onto a paint-soaked sponge, they were allowed to walk down the corridor, leaving hind footprints on the paper. In each walking track, three footprints were analyzed by a single observer, and the average of the measurements was used in SFI calculations. Several measurements were taken from the footprints: (I) distance between heel and third toe, print length (PL); (II) distance between the first and the fifth toe, the toe spread (TS); and (III) distance between the second and the fourth toe, the intermediary toe spread (ITS). All three measurements were taken from the right (experimental) sides. The SFI was calculated according to the following formula, as described by Bain et al. [15]:

$$SFI = -38.3\left(\frac{EPL - NPL}{NPL}\right) + 109.5\left(\frac{ETS - NTS}{NTS}\right) + 13.3\left(\frac{EIT - NIT}{NIT}\right) - 8.8$$

The SFI oscillates around 0 for normal nerve function, and a figure of around −100 indicates total dysfunction.

10.2.2.4 Wet Weight of Muscles

Following the electrophysiological examination 12 weeks after severing of the nerve, the right

Table 10.1 Compound muscle action potentials (CMPs) (mv)

	The triceps surae					Anterior tibialis	
	Pre-	Post-	4 Weeks	8 Weeks	12 Weeks	Pre-	Post-
Control	39.89±9.13	39.02±9.08	38.69±10.52	39.17±8.63	39.54±11.69	29.71±9.09	28.34±11.48
L4	40.23±10.34	39.56±10.51	38.36±9.42	38.91±10.28	39.20±11.15	30.12±8.19	27.89±8.34
L5	39.04±10.29	39.19±11.73	37.91±10.75	38.27±11.49	38.77±7.83	29.52±7.58	28.19±9.88
L6	39.35±9.68	39.21±11.94	39.13±11.43	39.77±9.41	39.68±9.37	30.34±10.13	28.46±7.22

Table 10.2 The mean latency (ms)

	The triceps surae					Anterior tibialis	
	Pre-	Post-	4 Weeks	8 Weeks	12 Weeks	Pre-	Post-
control	1.42±0.19	1.56±0.16	1.81±0.21	1.82±0.17	1.79±0.23	1.51±.24	1.53±0.21
L4	1.51±0.21	1.61±0.23	2.14±0.19	1.84±0.27	1.76±0.18	1.47±0.19	1.49±0.17
L5	1.49±0.17	1.53±0.17	2.26±0.32	1.79±0.13	1.74±0.21	1.53±0.23	1.46±0.28
L6	1.56±0.14	1.59±0.19	1.76±0.19	1.70±0.24	1.78±0.19	1.49±0.18	1.54±0.15

triceps surae, anterior tibialis and biceps femoris were dissected out from bilateral hind limbs and weighed while wet. After weighing, the muscles were fixed with 4 % paraformaldehyde in phosphate-buffered saline (PBS, pH 7.4) at 4 °C.

10.2.2.5 Electron Microscopic Observation of the Ultrastructure of the Target Muscle and Motor End-Plate (MEP)

The triceps surae, anterior tibialis and biceps femoris were double-fixed in 4 % glutaraldehyde and osmic acid, dehydrated with pyroracemic acid, and embedded in EPON812. They were then sliced using a LKBNOVA ultramicrotome (LKB Inc., Bromma, Sweden), double-dyed with lead and uranium and observed using a Philips CM120 transmission electron microscope (Philips Inc., Amsterdam, The Netherlands) for ultrastructure of the triceps surae, anterior tibialis and biceps femoris and MEP.

10.2.2.6 Histological Examination

The triceps surae, anterior tibialis and biceps femoris were fixed in 10 % formalin for 40 h, dehydrated and paraffin-embedded. They were then sliced along the middle of the muscle into 5 mm sections, hematoxylin and eosin stained, and analyzed using the FW4000 digital imaging workstation (Leica Inc., Solms, Germany).

10.2.2.7 Fibrotic Components in the Target Muscle

To observe the fibrotic level, the triceps surae, anterior tibialis and biceps femoris were dyed with Masson's trichrome. The sections were analyzed using the FW4000 digital imaging workstation (Leica Inc., Solms, Germany).

10.2.2.8 Statistics

Results were expressed as mean ± standard deviation, and SPSS 11.0 software (SPSS Inc., Chicago, IL) was used. One-way analysis of variance was used to determine significant differences between the groups, with the Student–Newman–Keuls test for multiple comparisons. Differences were considered significant at $P<0.05$. We certify that all applicable institutional and governmental regulations concerning the ethical use of animals were followed during the course of this research.

10.2.3 Results

The animals in all groups survived for the duration of the experiments, without wound infection or self-eating of limbs.

10.2.3.1 Compound Muscle Action Potentials (CMPs)

Before and after severing the nerve, the mean latencies of CMP and maximum CMP amplitudes in

			Biceps femoris				
4 Weeks	8 Weeks	12 Weeks	Pre-	Post-	4 Weeks	8 Weeks	12 Weeks
28.91±8.27	28.75±10.59	28.59±10.32	20.05±4.34	18.01±7.18	18.3±5.67	18.59±6.91	18.91±4.37
27.13±9.57	27.8±9.24	27.95±9.37	20.91±3.82	18.51±5.39	17.85±3.28	18.37±4.81	18.59±3.85
26.89±10.69	27.58±9.37	27.87±10.19	19.83±5.19	17.94±4.87	17.37±5.14	17.96±5.38	18.46±4.76
28.04±10.81	28.57±8.61	28.96±9.17	20.47±4.98	18.35±7.33	18.19±4.79	18.43±4.69	18.89±5.19

			Biceps femoris				
4 Weeks	8 Weeks	12 Weeks	Pre-	Post-	4 Weeks	8 Weeks	12 Weeks
1.78±0.24	1.77±0.16	1.79±0.13	1.54±0.23	1.53±0.27	1.69±0.28	1.75±0.24	1.83±0.27
2.05±0.19	1.84±0.24	1.81±0.27	1.49±0.18	1.51±0.15	2.01±0.12	1.83±0.19	1.81±0.17
2.13±0.26	1.86±0.17	1.83±0.16	1.57±0.26	1.56±0.17	2.17±1.17	1.87±0.23	1.83±0.24
1.75±0.29	1.74±0.27	1.72±0.28	1.49±0.24	1.54±0.26	1.77±0.21	1.82±0.17	1.76±0.17

Table 10.3 Sciatic functional index (SFI)

	Pre-operation	4 Weeks	8 Weeks	12 Weeks
Control	–5.28±2.39	–8.12±2.25	–7.56±3.32	–7.52±2.41
L4	–4.78±2.16	–10.17±3.59	–9.53±2.14	–8.27±2.49
L5	–5.19±3.07	–10.35±3.72	–9.87±2.81	–8.93±3.11
L6	–4.85±2.43	–9.25±3.06	–8.27±2.37	–7.82±2.79

groups A–D for the triceps surae, anterior tibialis and biceps femoris on the right side of rats ($n=10$) were, assessed respectively. There were no significant differences in either the mean latency or mean maximum CMP amplitude among the four groups. At the 4th week after surgery, there were significant differences in the mean latency among the four groups. Then at the 8th and 12 week after surgery, the differences in either the mean latency or mean maximum CMP amplitude among groups A-D for the triceps surae, anterior tibialis and biceps femoris on the left side of the rats ($n=10$) became insignificant (Tables 10.1 and 10.2).

10.2.3.2 Sciatic Functional Index (SFI)

Reproducible walking tracks could be measured from all rats. Before severing the nerve, the mean SFI was assessed in groups A, B, C and D, with no significant differences observed ($P>0.05$). At the 4th, 8th, and 12th week after surgery, the mean SFI was again assessed, with no significant differences found among the four groups at any of the time points ($P>0.05$) (Table 10.3).

10.2.3.3 Wet Weight of Muscles

The wet weight of the triceps surae (in grams) in groups A–D was 3.23±0.77, 2.91±0.84, 2.85±0.93, and 3.01±1.16 g, respectively. The wet weight of the anterior tibialis in groups A–D was 1.01±0.14, 0.85±0.22, 0.81±0.29 and 0.98±0.17 g, respectively. The wet weight of the biceps femoris in groups A–D was 5.26±1.13, 4.54±1.33, 4.28±1.46 and 5.13±1.59, respectively (Table 10.4). The results show that, for each of the different muscles, the mean muscle weight value of group A was always higher than that of groups B–D. But there was no significant difference in muscle weight between groups A, B, C and D ($P>0.05$).

10.2.3.4 Electron Microscopic Observation

In group A, organelles such as myofilaments and dense bodies were well organized in muscle cells of the triceps surae, anterior tibialis and biceps femoris. A similar result can be seen for groups B–D, except for 2 samples in groups B and 4 samples in C, in which degeneration of the muscle cells was observed (Figs. 10.3 and 10.4). The course of the myofilaments was malaligned and underwent vacuolar or globular changes. In group A, the normal structure of the MEP was seen between cells of the triceps surae, anterior tibialis and biceps femoris. A number of organelles including mitochondria and synaptic vesicles could be seen in the nerve endings. The ultrastructure of the MEP in groups B–D was similar to that in group A, without apparent degeneration (Fig. 10.5).

Table 10.4 Wet weight of muscles (g)

	The triceps surae	Anterior tibialis	Biceps femoris
Control	3.23±0.77	1.01±0.14	5.26±1.13
L4	2.91±0.84	0.85±0.22	4.54±1.33
L5	2.85±0.93	0.81±0.29	4.28±1.46
L6	3.01±1.16	0.98±0.17	5.13±1.59

10.2.3.5 Histological Examination

In group A, the muscle cells of the triceps surae, anterior tibialis and biceps femoris were well arranged and closely-contacted, with no infiltration of the connective tissue seen in muscle bundles. In groups B–D, the muscle cells were

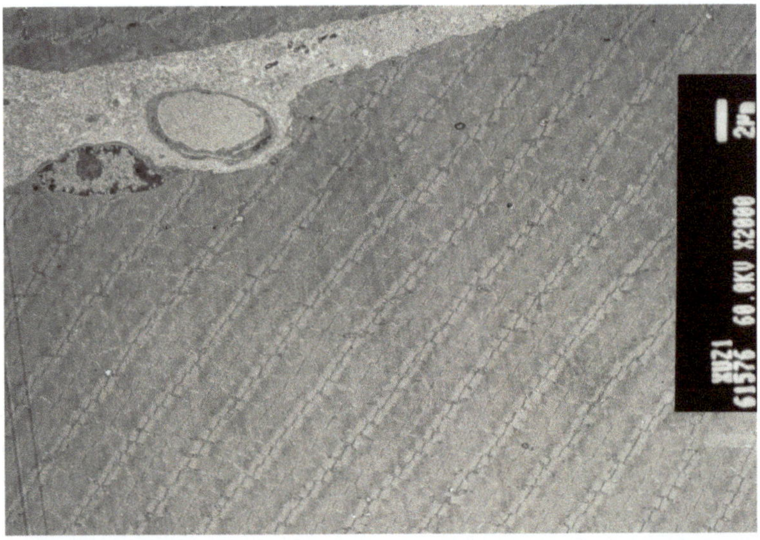

Fig. 10.3 In group A–D, organelles such as myofilaments and dense bodies were well organized in muscle cells (Electron microscopic × 2,000)

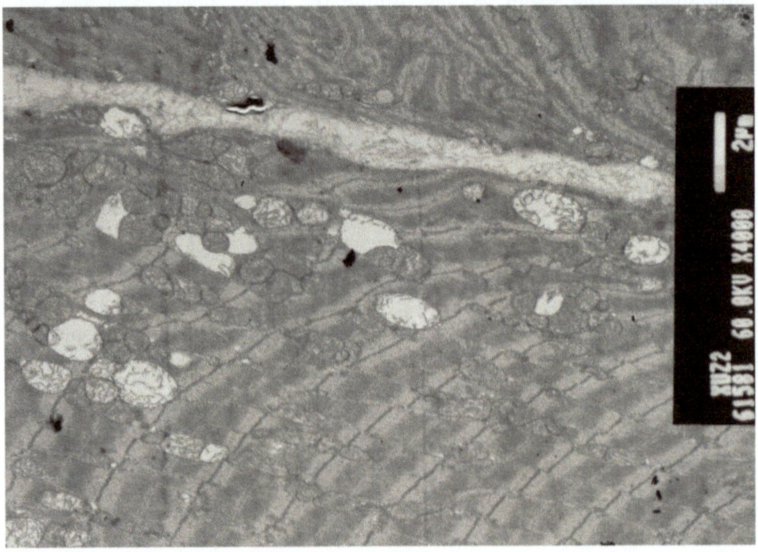

Fig. 10.4 A few samples in groups B and C, in which degeneration of the muscle cells was observed (Electron microscopic × 4,000)

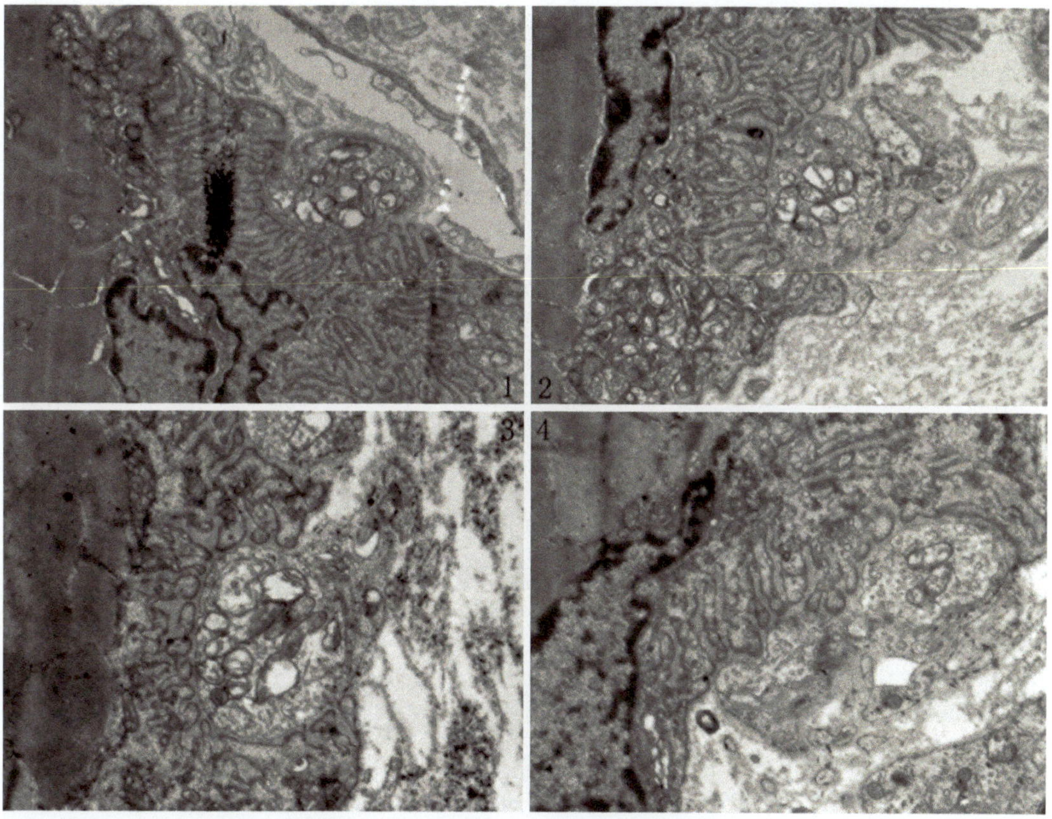

Fig. 10.5 In group A–D, the normal structure of the MEP (1.2×5,000; 3.4×10 K)

similarly arranged, except for 3 samples in B and 5 samples in C, in which the muscle cells were loosely arranged, with a rounded shape and smaller area compared with the control group, indicating degeneration of the muscle cells (Figs. 10.6, 10.7, and 10.8).

10.2.3.6 Analysis of Fiber Components

In group A, the triceps surae, anterior tibialis and biceps femoris cells were funicular and well arranged, and small amounts of collagen could be seen between the muscle bundles. Similar results were found for groups B–D, except for 2 samples in groups B and 4 samples in C (Figs. 10.9, 10.10, and 10.11), in which fibrosis of the muscle became worse and the cells became smaller than those in group A. Large numbers of collagen fibers were seen between the muscle bundles, with collagen infiltration. The percentage of connective tissue in the triceps surae in groups A–D was 12.58±1.45, 13.47±2.32, 13.24±2.76 and 13.30±2.56, respectively. The percentage of connective tissue in the anterior tibialis in groups A–D was 12.23±2.15, 12.98±2.47, 13.29±3.66 and 13.14±1.46 respectively. The percentage of connective tissue in the biceps femoris in groups A–D was 13.08±1.09, 13.57±2.52, 13.74±1.96 and 13.58±2.31respectively. The difference between group A and groups B, C, D was not significant ($P > 0.05$) (Table 10.5).

10.2.4 Discussion

Finding a way to solve the problem of neurogenic bladder would not only improve the quality of life in patients who have suffered from SCI, but also improve the rate of mortality. For spastic bladder, the Brindley Sacral Anterior Root Stimulator has proven to be an effective solution. Artificial bladder micturition reflex has also been demonstrated to be useful, in both animal and

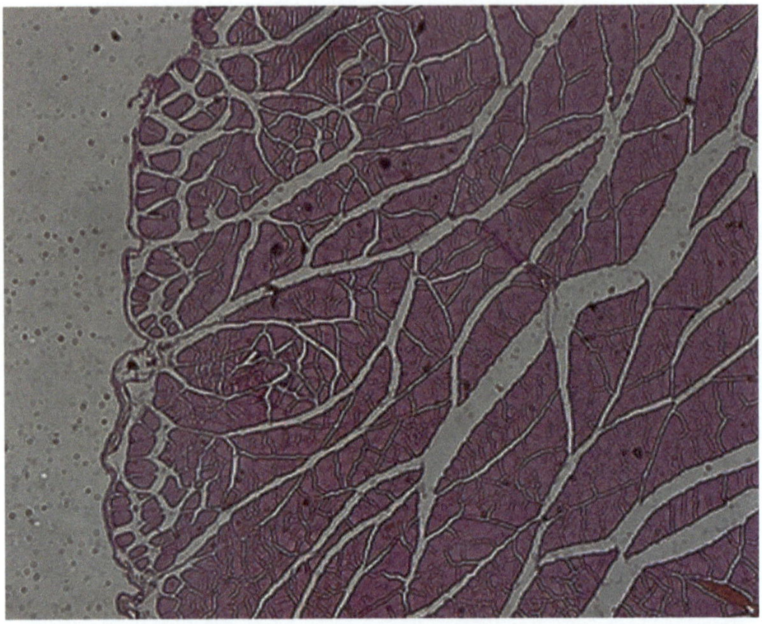

Fig. 10.6 In group A, the muscle cells were well arranged and closely-contacted (Hematoxylin and eosin stained × 100)

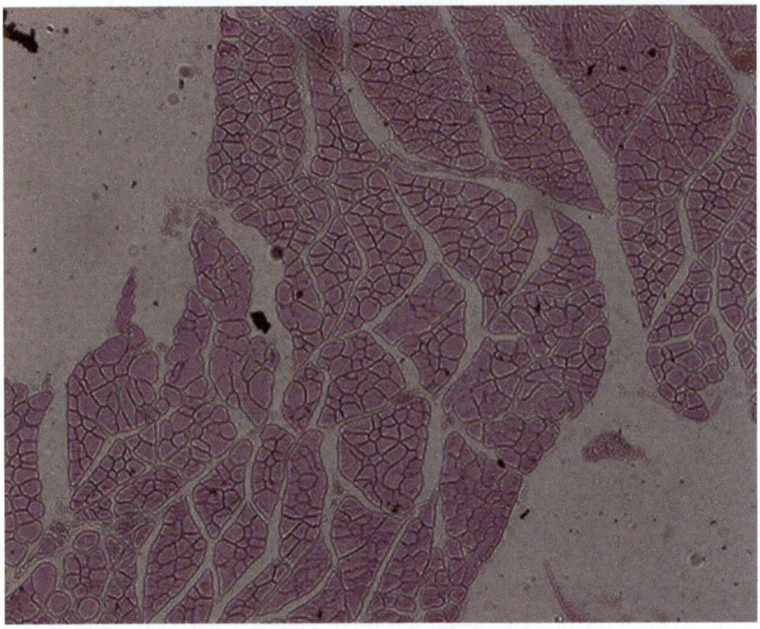

Fig. 10.7 In groups B–D, the muscle cells were similarly arranged (Hematoxylin and eosin stained × 100)

human studies. For the atonic bladder, using the normal abdominal reflex above the level of injury to reconstruct the bladder reflex arc has been performed clinically [12]. However, this requires a sural nerve transplant, which is nearly 30 cm long, to bridge the T_{11} and S_2 nerve root. Thus for patients with atonic bladder caused by conus medullaris injury, and whose motor function of the lower extremities is normal, using the normal S_1 nerve root as the donor nerve for the bladder to reconstruct the artificial bladder reflex arc is apparently advantageous over using the abdominal reflex, in that the distance of nerve regeneration required is about 30 cm less. In addition, there is no need for nerve transplantation, and instead a direct nerve root end-end anastomosis is

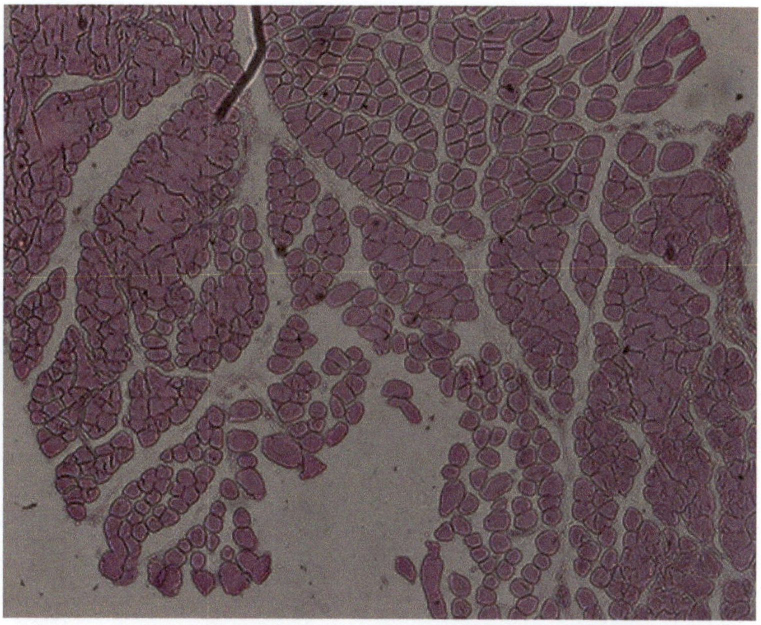

Fig. 10.8 A few samples in groups B and C, degeneration of the muscle cells (Hematoxylin and eosin stained × 100)

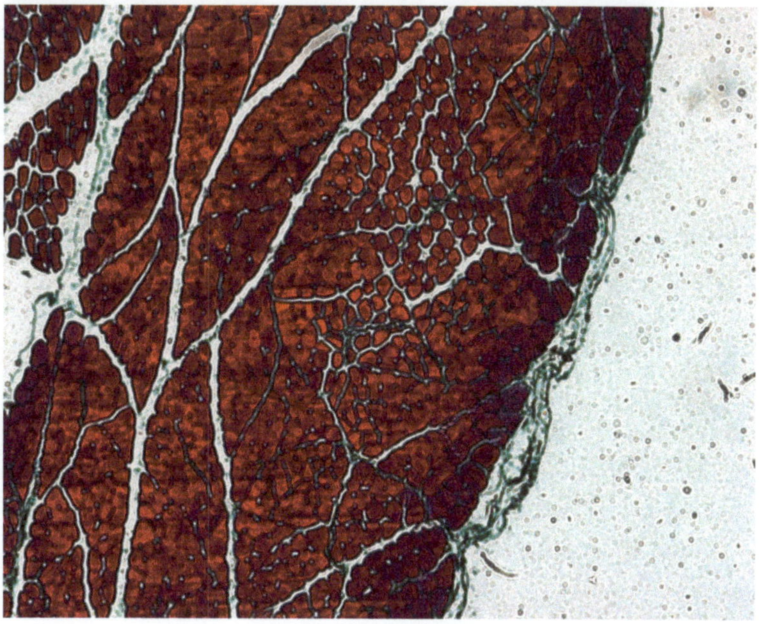

Fig. 10.9 In group A, small amounts of collagen could be seen between the muscle bundles (Masson's trichrome stained × 100)

required. Furthermore, the S_1 anterior root is large in diameter with abundant nerve fibers, which is favorable for functional recovery of the bladder after nerve regeneration. Anatomically, the S_1 nerve root runs downward and primarily participates in the functioning of the sciatic nerve, which is mainly composed of L_4, L_5, S_1, S_2 and S_3. In his study of repairing brachial plexus injury by using the C_7 nerve root of the healthy side, GU et al. confirmed that severing a single cervical nerve root did not result in significant long-term effects on the motor and sensory functions of the upper extremities, owing to compensatory action between the nerve roots [16, 17]. The main functions of the lower extremities are weight-bearing and walking, without needing

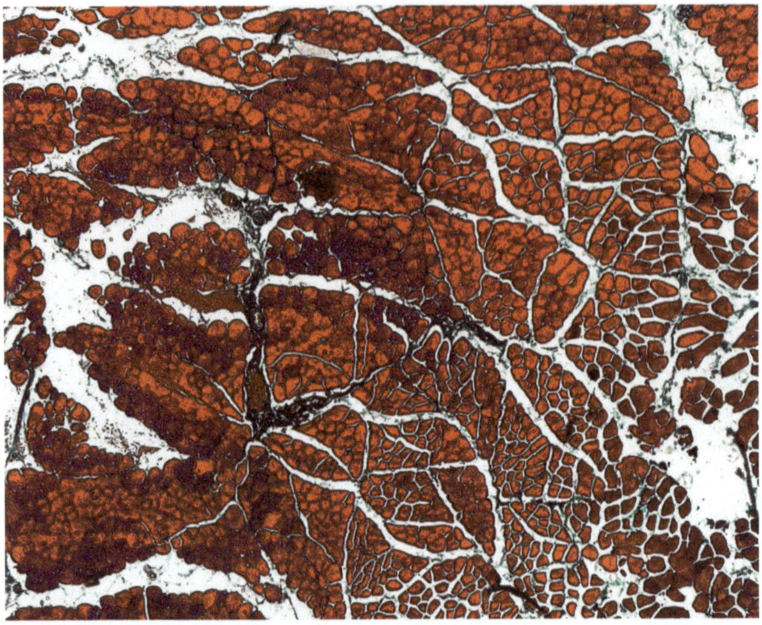

Fig. 10.10 In groups B–D, the muscle cells were similarly arranged (Masson's trichrome stained × 100)

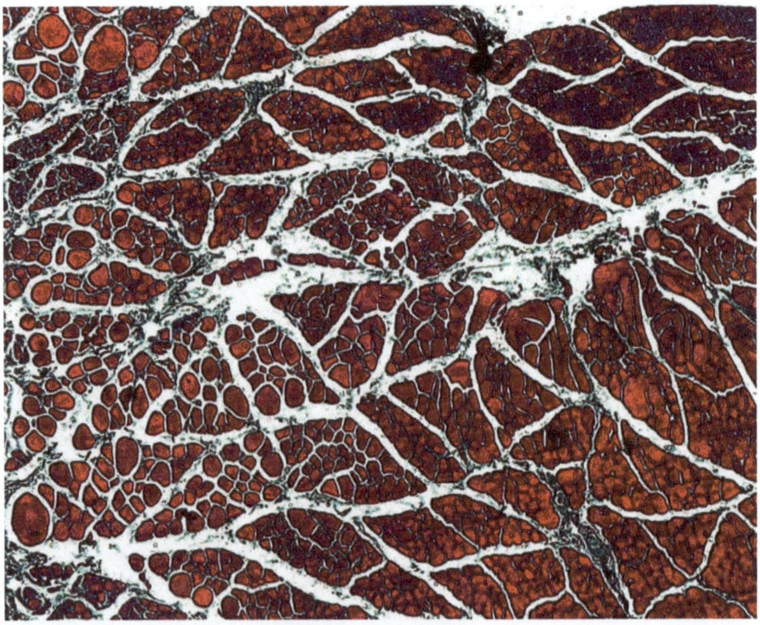

Fig. 10.11 A few samples in groups B and C, degeneration of the muscle cells (Masson's trichrome stained × 100)

Table 10.5 The percentage of connective tissue (%)

	The triceps surae	Anterior tibialis	Biceps femoris
Control	12.58 ± 1.45	12.23 ± 2.15	13.08 ± 1.09
L4	13.47 ± 2.32	12.98 ± 2.47	13.57 ± 2.52
L5	13.24 ± 2.76	13.29 ± 3.66	13.74 ± 1.96
L6	13.30 ± 2.56	13.14 ± 1.46	13.58 ± 2.31

precise innervation. In addition, the distribution of these functions is also relatively compact in the cerebral cortex, and therefore mutual domination and compensation between nerve roots is probably more apparent.

In this study, we investigated the disadvantages of an operation that uses a normal sacral plexus nerve root as the dynamic nerve to re-establish micturition of the bladder. Considering that the sacral plexus in SD rats is predominantly composed of L4, L5 and L6 nerve roots, we observed the effects on lower limb function when L4, L5 and L6 nerve roots, respectively, were severed [14]. From the EMG results, we found that there was an observable effect at 4 weeks after severance of L4 or L5, compared with normal controls, and these effects were decreased at 8 and 12 weeks after L4 or L5 was severed. However, there was no significant effect produced by severance of L6.

From the SFI results, there was a small decrease at 4 weeks after each operation, compared with the normal controls, but the change was not significant, and the effects decreased over time. A similar result was found for the wet weight of muscles, except for several samples. By electron microscopy, histological examination and analysis of fiber components, we also found that significant degeneration occurred in only a few muscle samples.

In conclusion, severing of one of the normal sacral plexus nerve roots in SD rats was found to be feasible, and did not obviously impact on lower limb motor function. The single normal sacral plexus nerve root can be used as a dynamic nerve to re-establish micturition of the bladder, and may possibly be feasible to help repair other function failure ascribe to peripheral nerve injury, such as motor function disorder of the lower limbs. As there may be species-specific differences between human beings and rats, more studies are required to make the operation safer and more feasible. The mechanism of the compensation observed in this operation also remains to be elucidated in future studies.

10.3 Clinical Application

10.3.1 Introduction

Neurogenic bladder in patients with spinal cord injury (SCI) poses a challenge to researchers and clinicians alike [18–21]. Among the treatments reported, one of the methods used to treat spastic bladder caused by SCI above the level of the conus medullaris is electric stimulation [22]. Another method involves the use of healthy tendon reflex mechanisms below the paraplegic level. We have demonstrated in a previous work that the Achilles tendon reflex can be used in humans to successfully reconstruct spastic bladder function in patients with complete suprasacral SCI; We anastomosed the S1 ventral root (VR) with the S2 and/or S3 VRs to reconstruct the bladder reflex arc [9]. With regard to atonic bladders caused by conus medullaris injuries, it has been shown that reconstructing the innervation to the bladder above the level of SCI can produce urination by abdomen-to-bladder reflex contractions [11, 12]. However, this procedure requires a sural nerve transplant to bridge the T10 (T11) and S2/3 nerve roots, which can be up to 30 cm long.

We think that this procedure can be avoided in some patients with atonic bladder due to a low conus medullaris (S2–5) injury, such as from an L1 fracture, in whom bladder dysfunction is present but the majority or all of the motor functions of the lower extremities remain intact. In the case of such patients, the S1 ventral root of the unaffected spinal segment could be used to re-innervate the atonic bladder. However, severing the S1 nerve root could cause further functional losses in the lower limbs.

The brachial and sacral plexuses are organized in a similar manner, and Gu et al. [16] have confirmed that the ipsilateral C7 root could be used to repair brachial plexus injuries, and that severing C7 did not produce significant long-term effects on motor and sensory functions of the upper extremities, probably due to the compensatory action of the other nerve roots. Therefore, hypothetically, if the S1 root is divided, other nerves should be able to compensate for damaged nerves and control muscles that were originally controlled by S1. In order to examine this, in our previous experiments, we confirmed that division of the L6 nerve root in the monkey sacral plexus, which is analogous to the S1 root in humans, did not permanently affect limb function [23]. Based on the results of these preclinical experiments, this paper describes our attempts to re-innervate the atonic bladder of patients who had previous conus medullaris injuries, by using an ipsilateral S1 root transfer.

10.3.2 Indications

Patients who presented with bladder atonia after conus medullaris injury and in whom motor function of the lower extremities was preserved.

10.3.3 Surgical Technique

The surgical procedure was performed under general anesthesia. A standard laminectomy was performed from L5 to S3 with the patient lying in the prone position. The dura was opened through a paramedian incision, which exposed the dorsal and ventral roots of the S1–S4 nerves. The S1 nerve root was located using the L5/S1 intervertebral space as a marker; the S2–4 nerve roots were located in descending order. The ventral and dorsal roots at the dural incision were identified according to their anatomical characteristics. The VRs of S1, S2 and S3 were identified and separated from their respective dorsal roots by microdissection. An electric stimulator was used to stimulate the S1 VR to observe muscle contractions of the lower limb in order to verify that the root was indeed the S1 root. If the function of the S1 root was normal, the unilateral S1 VR and the S2/3 VRs on the same side were transected using microsurgery and anastomosed with 9–0 non-absorbable sutures (Figs. 10.12 and 10.13). The wound was sutured in three layers with an external drain.

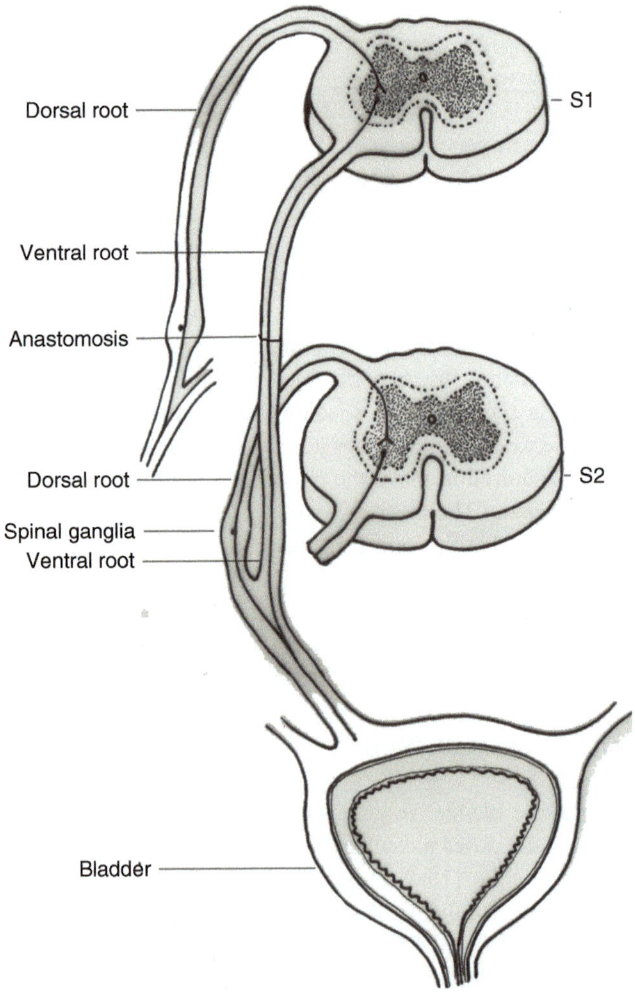

Fig. 10.12 Creation of microsurgical anastomoses between the S1 and S2/3 ventral roots

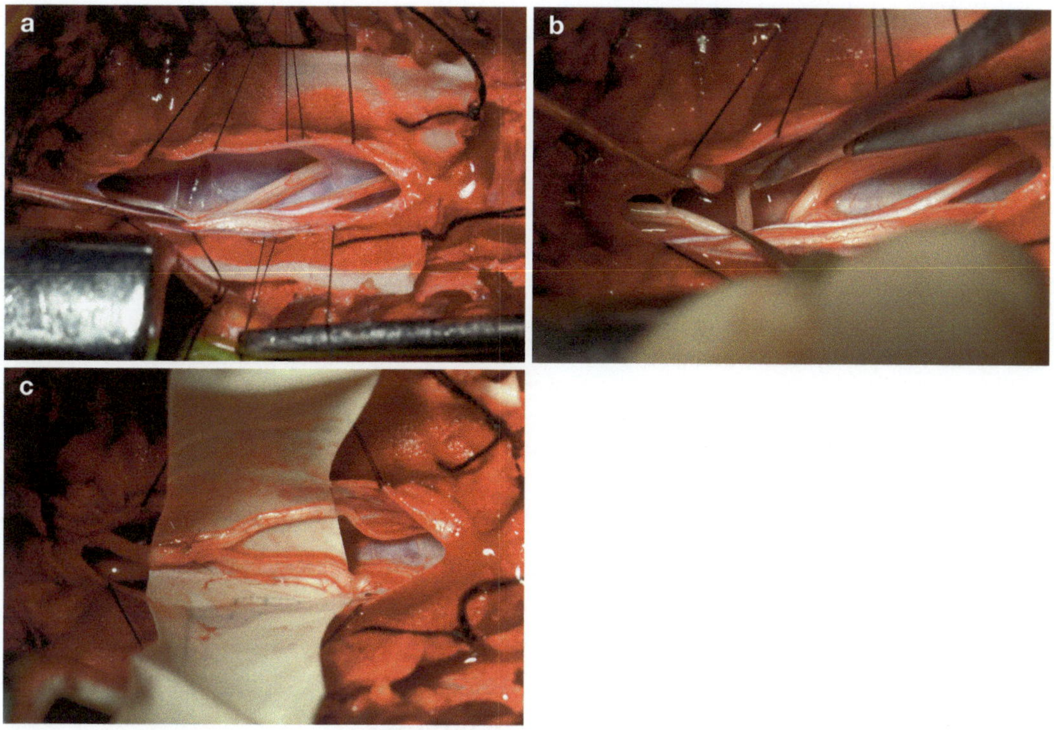

Fig. 10.13 Surgical procedure for reconstructing the bladder reflex pathway in a 25-year-old man. (**a**) Standard laminectomies from L5 to S3 were performed with the patient lying in the prone position. The dura was opened through a paramedian incision, exposing the dorsal and ventral roots of the S1, S2 and S3 nerves. (**b**) The ventral roots of S1, S2 and S3 were identified and separated from their respective dorsal roots by microdissection. (**c**) The S1 and S2/3 ventral roots were anastomosed

10.3.4 Case Report

A 35-year-old man was referred for evaluation and treatment of atonic bladder caused by a conus medullaris injury in March 2008. He had fallen from an elevated height, which resulted in a compression fracture of the L2 vertebra. The patient presented with bladder atonia that was associated with the conus medullaris injury, and motor function of the lower extremities was preserved. Physical examination showed that the muscle strength of both lower extremities was practically normal. Both spontaneous voiding and sensation of bladder fullness were absent, and cystometrography revealed that the patient had an areflexic bladder. The patient urinated via a bladder stoma, and there was no residual urine in the bladder. Surgical reconstruction of the bladder reflex pathway was scheduled and performed 6 months after the injury. The right S1 VR and the S2/3 VRs on the same side were transected using microsurgery and anastomosed.

On the second day and 1, 3, 6, and 12 months after the operation, the strength of the muscles innervated by the S1 nerve was examined, including the gastrocnemius, soleus, abductor hallucis and extensor digitorum brevis. Muscle strength was evaluated using the MRC grading scale. At 1, 3, 6, 12, 18, 24 and 36 months after the operation, urodynamic tests were conducted a 4-channel urodynamic system (Dantec, Denmark), to evaluate urine volume, maximum detrusor pressure, maximum flow and residual urine volume.

On the second day after the operation, the power of the S1-innervated muscles decreased by 1 grade (MRC grade) in comparison with the preoperative levels. However, at 3 months after the operation, muscles innervated by the S1 nerve root had recovered their preoperative strength levels.

The patient experienced recovery of bladder storage and voiding functions, and no nocturnal urinary incontinence was noted at postoperative 12 months. The patient was also able to void voluntarily. Interestingly, he also regained bladder sensory function, which was characterized by the ability to sense a full bladder and perceive the desire to void. When the pre- and postoperative urodynamic study results were compared, we found significant improvement in bladder function by the 12th postoperative month.

10.3.5 Discussion

In the nine patients included in this study whose S1 nerve root was severed, physical examination revealed that postoperative muscle strength decreased slightly, and recovered 3 months postoperatively. These results indicate that the unfavorable effect of severing the S1 root can be compensated for by other nerve roots; therefore, it may be safe to transfer the S1 root to re-innervate an atonic bladder [24].

The procedure we used allowed patients to initiate voiding voluntarily by stimulating the Achilles tendon, which is mainly innervated by the S1 nerve root, via percussion of the Achilles tendon. To our surprise, some patients who regained bladder storage and emptying functions also regained bladder sensation, which was demonstrated by their ability to sense a full bladder and have the desire to void. All of these patients were able to void voluntarily. We believe that it is reasonable that patients would gain conscious control over their bladder and external urethral sphincter as both were postoperatively innervated by the S1 segment, which is above the level of injury. Furthermore, as only the ventral roots were involved in the anastomosis, regained sensation could only result from regenerating sensory fibers traveling in the ventral root, or a modulation/enhancement of intact S1 or thoraco-lumbar afferents resulting from re-innervation of the muscle itself. Future studies are required to evaluate the mechanism underlying the reacquisition of urinary function as well as the time course of reflex arc generation.

In addition to the favorable results, this novel approach allows for tension-free nerve anastomosis without the need for nerve transplantation, since the S1 and S2/3 nerve roots overlap in the cauda equina; it also shortens the required duration of nerve regeneration, as the anastomotic site is relatively low. Furthermore, the S1 VR is relatively large in diameter with abundant nerve fibers, which is favorable for functional recovery of the reinnervated bladder. Finally, the anastomosis of simple motor nerves avoids axonal mismatch, which ensures that nerve function can recover smoothly.

Therefore, in patients with atonic bladder caused by SCI whose motor function of the lower extremities is normal, the use of the normal S1 root as the donor nerve to re-innervate the atonic bladder is advantageous over using the abdomen-to-bladder reflex.

References

1. Tanagho EA, Schmidt RA. Electrical stimulation in the clinical management of the neurogenic bladder. J Urol. 1988;140:1331–9.
2. Selzman AA, Hampel N. Urologic complications of spinal cord injury. Urol Clin North Am. 1993;20: 453–64.
3. Mitchell ME, Kulb TB, Backes DJ. Intestinocystoplasty in combination with clean intermittent catheterization in the management of vesical dysfunction. J Urol. 1986;136:288–91.
4. Brindley GS, Polkey CE, Rushton DN, Cardozo L. Sacral anterior root stimulators for bladder control in paraplegia: the first 50 cases. J Neurol Neurosurg Psychiatry. 1986;49:1104–14.
5. Xiao CG, Godec CJ. A possible new reflex pathway for micturition after spinal cord injury. Paraplegia. 1994;32:300–7.
6. Xiao CG, de Groat WC, Godec CJ, Dai C, Xiao Q. "Skin- CNS-bladder" reflex pathway for micturition after spinal cord injury and its underlying mechanisms. J Urol. 1999;162:936–42.
7. Xiao CG, DuMX DC, Li B, Nitti VW, de Groat WC. An artificial somatic- central nervous system-autonomic reflex pathway for controllable micturition after spinal cord injury: preliminary results in 15 patients. J Urol. 2003;170:1237–41.
8. Zheng XY, Hou CL, Zhong HB, Xu RS, Chen AM, Xu Z, Wang JH. Reconstructed bladder innervation below the level of spinal cord injury: the knee-tendon to bladder artificial reflex arc. J Spinal Cord Med. 2009;32:79–85.

9. Lin H, Hou C, Zhen X, Xu Z. Clinical study of reconstructed bladder innervation below the level of spinal cord injury to produce urination by Achilles tendon–to-bladder reflex contractions. J Neurosurg Spine. 2009;10:452–7.
10. Zheng X, Hou C, Chen A, Li J, Xu Z, Wang J. Optimal timing of operation for repairing atonic bladder after medullary cone injury: an experimental study in rats. Spinal Cord. 2008;46(8):574–81.
11. Wang JW, Hou CL, Lu N. Long-term functional study on the artificial bladder reflex arc established by abdominal reflex after spinal cord injury. Chin J Microsurg. 2002;25:280–4.
12. Lin H, Hou C, Zhong G, Xie Q, Wang S. Reconstruction of reflex pathways to the atonic bladder after conus medullaris injury: preliminary clinical results. Microsurgery. 2008;28:429–35.
13. Lin H, Hou C, Zhen X. Bypassing spinal cord injury: surgical reconstruction of afferent and efferent pathways to the urinary bladder after conus medullaris injury in a rat model. J Reconstr Microsurg. 2008;24:575–82.
14. Fumio A, Butler M, Blomberg H, et al. Variation in rat sciatic nerve anatomy: implications for a rat model of neuropathic pain. J Peripher Nerv Syst. 2000;5:19–21.
15. Bain J, Mackinnon S, Hunter DA. Functional evaluation of complete sciatic, peroneal, and posterior tibial nerve lesions in the rat. Plast Reconstr Surg. 1989;83:129–38.
16. Gu Y, Xu J, Chen L, Wang H, Hu S. Long term outcome of contralateral C7 transfer: a report of 32 cases. Chin Med J (Engl). 2002;115:866–8.
17. Xu L, Gu Y, Xu J, Lin S, Chen L, Lu J. Contralateral C7 transfer via the prespinal and retropharyngeal route to repair brachial plexus root avulsion: a preliminary report. Neurosurgery. 2008;63:553–8.
18. Chuang DC, Chang PL, Cheng SY. Root reconstruction for bladder reinnervation: an experimental study in rats. Microsurgery. 1991;12:237–45.
19. Carlsson CA, Sundin T. Reconstruction of efferent pathway to the urinary bladder in a paraplegic child. Rev Surg. 1967;24:73–6.
20. Fujishiro T, Takahashi S, Enomoto H, Ugawa Y, Ueno S, Kitamura T. Magnetic stimulation of the sacral roots for the treatment of urinary frequency and urge incontinence: an investigational study and placebo controlled trial. J Urol. 2002;168:1036–9.
21. Gasparini ME, Schmidt RT, Tanagho EA. Selective sacral rhizotomy of the reflex neuropathyic bladder: a report on 17 patients with long-term follow-up. J Urol. 1992;148:1207–10.
22. Brindley GS. The first 500 patients with sacral anterior root stimulator implants: general description. Paraplegia. 1994;32:795–805.
23. Lin H, Xu Z, Liu Y, Chen A, Hou C. The effect of severing L6 nerve root of the sacral plexus on lower extremity function: an experimental study in rhesus monkeys. Neurosurgery. 2012;70:170–7.
24. Lin H, Hou C. Transfer of normal S1 nerve root to reinnervate atonic bladder due to conus medullaris injury. Muscle Nerve. 2013;47:241–5.

Electrical Stimulated Micturition: Sacral Anterior Root Stimulator + Sacral Deafferentation

11

Shimin Chang and Chunlin Hou

The urinary system plays three major functions in the human body: (1) the kidneys filter waste products from the body to make urine; (2) the urine then moves through the ureters to the bladder which stores the urine; (3) the urine is then released from the body through the urethra.

Spinal cord injury can affect bladder control. This is because the nerves responsible for communicating to and from the brain to indicate when the bladder is full are located at the terminal base of the spinal cord (the terminal cord cone).

Bladder control with spinal cord injury can be a difficult but manageable situation. The Finetech-Brindley Sacral Anterior Root Stimulator (SARS or Vocare, an implantable neuroprosthesis) is an established and successful device for bladder and bowel emptying in Spinal Cord Injury, and is accompanied by a rhizotomy of the posterior (sensory) sacral roots [1, 2]. The Brindley procedure consists of a stimulator for sacral anterior root stimulation and a rhizotomy of the dorsal sacral roots to abolish neurogenic detrusor overactivity. Stimulation of the sacral anterior roots enables controlled micturition, defecation, and erections, while dorsal root rhizotomy (sacral de-afferatation) enables a good reservoir function [3, 4]. The Brindley procedure (SARS + SDAF) has been used in more than 3,000 patients [5].

S. Chang
Department of Orthopedic Surgery, Yangpu Hospital, Tongji University School of Medicine, Shanghai, People's Republic of China
e-mail: shiminchang11@aliyun.com

C. Hou, M.D. (✉)
Department of Orthopedic Surgery, Changzheng Hospital, The Second Military Medical University, Fengyang Road 415, Shanghai 200003, People's Republic of China
e-mail: chunlin_hou@163.com

11.1 The Brindley Stimulator Device (An Implantable Neuroprosthesis)

The FinetechBrindley Bladder Control System (branded VOCARE neuroprosthesis) is a radiofrequency powered motor control neuroprosthesis, which is composed of an external and implanted part [1, 2]. The implanted part consists of electrodes, connecting cables, and a receiver block. An external wireless device is used to control the implant and its function. Patients have to position an external stimulating device on the skin over the implanted receiver to evoke stimuli. The receiver does not have a battery. Electrical stimuli are evoked by radiofrequency waves. With the availability of separate stimulation of the sacral levels and various stimulation settings, it is possible to set various stimulation programs to optimize micturition, defecation, and penile erections.

A tripolar electrode cuff is used for intradural stimulation of the sacral anterior roots. A three-channel implant is composed of two books. The

C. Hou (ed.), *Functional Bladder Reconstruction Following Spinal Cord Injury via Neural Approaches*, DOI 10.1007/978-94-007-7766-8_11, © Springer Science+Business Media Dordrecht 2014

upper book contains three parallel slots for S3 (one slot) and S2 roots (two slots at one channel), and the lower contains one slot for S4 roots. Each slot contains one cathode in the centre and an anode at each of the two ends to avoid stimulation of tissue structures outside the slot. The two-channel implant allows stimulation of two root levels or sets of root levels. The four-channel implant has the same configuration as the three-channel implant but allows independent stimulation of four sets of roots. The choice for the number of channels depends on the number of different rootlet combinations that have to be stimulated. Each channel is connected to the subcutaneous receiver block by a silicone-coated cable.

Extradural electrodes are used in patients in whom intradural electrodes could not be placed due to, for example, arachnoiditis or a previous intradural electrode implantation that failed. Some medical centers and surgeons prefer to use extradural electrodes primarily for nearly all patients. The extradural implant has three helical electrodes at its end, which are also configured with a cathode between two anodes.

The Finetech-Brindley Bladder Control System (VOCARE) uses Functional Electrical Stimulation (**FES**) to stimulate the nerves which connect the spinal cord to the bladder and bowel. This allows patients with complete spinal cord injuries to empty their bladder when they decide to ("on demand"). It works in the following steps:

1. The user selects one of the stimulation programs.
2. The external controller (A) then sends power and control signals to the transmitter block (C) via the transmitter lead (B).
3. The user holds the transmitter block over the site of the implant receiver (D).
4. The implant receiver converts the signals from the transmitter block into electrical impulses.
5. The impulses travel via extremely flexible cables (E) to the electrodes (F).
6. The electrodes transfer the impulses to the nerves causing the appropriate muscles to contract at the appropriate time.

11.2 The Poststimulus Voiding Mechanism

Most of the small diameter parasympathetic efferent nerve fibres for innervation of the bladder are located in the sacral anterior roots (S2–S4/5). Small-diameter nerve fibres need a higher stimulus for their excitation than large-diameter fibres. Consequently, electrical stimulation of the anterior roots for detrusor contractions also causes contraction of the urethral sphincter due to stimulation of somatic large-diameter nerve fibres. This prevents emptying of the bladder. To overcome this problem, poststimulus voiding is used. The time to relax of striated muscles of the urethral sphincter is shorter than the relaxation time of smooth muscles of the detrusor. When intermittent stimulation pulse trains are applied, the difference in muscle relaxation time can be used to achieve a sustained detrusor muscle contraction with intervals of urethral sphincter relaxation (Fig. 11.1). These intervals in between stimulations allow a decrease of the urethral sphincter pressure while a high intravesical pressure remains. This results in poststimulus voiding with an intermittent pattern of the micturition

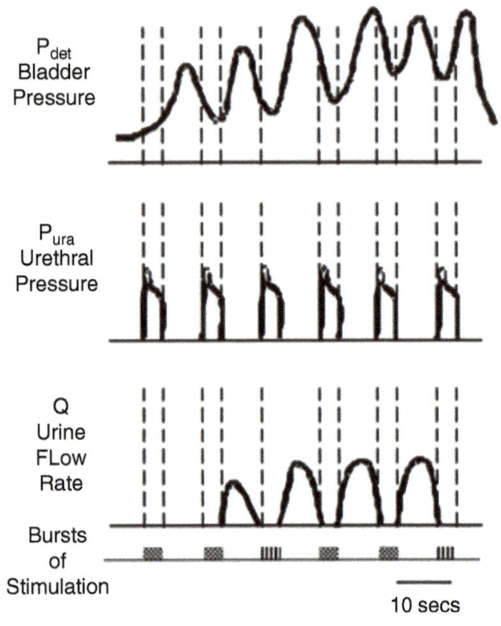

Fig. 11.1 Showing the post-stimulus voiding mechanism

flow [6]. A comparable mechanism has been used for defecation.

11.3 Dorsal Rhizotomy of the Sacral Nerves

Sauerwein in 1988 structurally expanded SARS with a sacral de-afferentation (SDAF), i.e. dorsal rhizotomy of sacral roots S2 till S5 [3]. A dorsal rhizotomy is important because it suppresses neurogenic detrusor overactivity and detrusor-external sphincter dyssynergia. This results in a low-pressure bladder without reflex contractions of the detrusor and subsequently continence. Moreover, it reduces autonomic dysreflexia [7, 8]. Therefore, a dorsal rhizotomy can also be applied in combination with intermittent catheterization to empty the bladder without implantation of a Brindley stimulator.

11.4 Indications and Contraindications

There are often two types that can characterize the bladder in spinal cord injuries: (1) a flaccid (floppy) bladder, and (2) a hyperactive (reflex) bladder.

The flaccid bladder can be a result of damage to the T12/L1 section and below. For a flaccid bladder, no trigger is detected by the nerves responsible for bladder control and therefore no message reaches the brain. The result is that the bladder can be filled in large volume (good storing function), but cannot be discharged spontaneously. When the urine collected is over the capacity of an adult's bladder (400 cm^3), the urine will back up the ureters and towards the kidneys. The reflux of the urine up the ureters can lead to kidney failure as well as complications, such as urinary tract infection, bladder stone formation, etc. Clinically, one of the more practical forms of management for a flaccid bladder is ICP.

A reflex bladder will be caused when damage to the spine is above the T12 level. For a reflex bladder, the spinal cone still has innervations to the bladder, but lose the upper control from the brain. It is characterized by detrusor hyperreflexia (cause incontinence), detrusor-external sphincter dyssynergia, high bladder pressures and, vesicoureteric reflux leading to renal impairment. The patient can pass urine at anytime and anywhere, which can be rather disconcerting.

Spinal cord injured patients with a suprasacral lesion usually develop a spastic bladder. The neurogenic detrusor overactivity and the overactive external sphincter cause incontinence and threaten these patients with recurrent urinary tract infections (UTI), renal failure and autonomic dysreflexia. All of these severe disturbances may be well managed by sacral deafferentation (SDAF) and implantation of a sacral anterior root stimulator (SARS).

Candidates criteria for patient selection including the following [9]:

1. Skeletally mature.
2. Intact reflex bladder contractions, i.e. sufficient efferent nerve pathways to the bladder and the bladder is able to contract. Contractions of at least 50 cm H_2O in males, or 30 cm H_2O in females need to be present during filling cystometry.
3. At least 12 months post-injury to ensure stable lesion.
4. A clinically complete spinal cord lesion.
5. Preoperative MRI excluding arachnoiditis at the level of the conus and cauda equine.
6. Good support network.

The contraindications including: (1) Poor or inadequate bladder reflexes, (2) Active or recurrent pressure ulcers, (3) Active sepsis (blood poisoning), (4) having an implanted cardiac pacemaker.

11.5 Surgical Technique

The patient is placed in prone position. The total operation consists two parts: electrodes implantation and dorsal root rhizotomy, and may involve in three or four incisions. Implantation can be performed in introdural or extradural at the sacral root level, deafferentation can be performed at the introdural at the sacral level, or at the conus medullaris at T12-L1 vartebral level (Table 11.1) [10–19].

Table 11.1 Comparison of different surgical procedures

Methods	Skin incision	SARS	SDAF
Brindley	L4-S2	Intradural	Introdural
Barcelona	L5-S3, T12-L2	Extradural	Conus
Tanagho	L4-S2	Intradural	Posterior ganglion
	L4-S2	Extradual	Introdural

1. **Intradural implantation:**

   A laminectomy from L3-L4 to S1-S2 is done for an intraduralrhizotomy and intradural implantation of the electrode cuff [1]. The dura and arachnoid are opened at midline to expose the sacral nerve roots. The anterior and dorsal components of the roots, especially relevant anterior roots for micturition, can be identified intradurally by electrical stimulation of these components while monitoring the effects on detrusor activity, blood pressure, and somatomotor responses. A rhizotomy of the identified dorsal sacral roots is done. The anterior sacral roots are positioned into the electrode cuff. The connecting cables are subcutaneously tunnelled to a subcutaneous pocket (lateral thoracic) for the receiver.

2. **Extradural implantation:**

   Implantation of extradural electrodes requires a laminectomy from L5-S1 to S3-S4 [15]. The dorsal rhizotomy is done at the level of the posterior ganglia of S2-S5. Electrical stimulation tests are used to identify the anterior and dorsal components of the sacral roots. The extradural electrode is implanted and fixated to the nerve using a strip of silicone rubber sheet which is sewn to itself and surrounds the nerve. The connecting cables and the receiver are implanted the same way as the intradural procedure.

3. **Surgical Techniques of Barcelona procedure**

   As a routine, we prefer extradural electrode implantation and SDAF at conusmedullaris (the Barcelona procedure) for our patients, and this will be described in detail [18]. The advantages of SDAF at conusmedullaris is less risk of damaging anterior roots, and almost no risk of failing to completely cut the posterior root. Its disadvantage is two skin incisions and requires more time to removal bone (laminectomy).

   The operation usually takes 4–5 h.

   If the anterolateral or lateral thoracic site is chosen for the receiver, the whole opertation can be done with the patient prone. If the abdominal or anterior thoracic site is chosen, the patient must be turned to supine for implantation of the receiver block.

   Premedication with atropine should be avoided, and if the patient has been taking an antimuscarinic drug, this should have been stopped at least 7 days before the operation. Neuromuscular blocker, if used during operation, should be stopped at the stage of testing the roots by stimulation.

   A longitudinal 10 cm skin incision is made from lower lumbar to sacral, centered at S1. Laminectomy from L5 to S3 vertebrae is done, exposing the bilateral nerve roots, including S2, S3 and S4. It is first identified by anatomic locations (L5 nerve root located in L5/S1 vertebrae), and then demonstrated by intraoperative electrical stimulation and observation of the lower leg muscle response (Table 11.2) and bladder pressure uprising (filled with 100 saline water) through indwelling balloon catheter.

   The parasympathetic autonomic efferents from sacral spinal cord to bladder detrusor are generally considered lying in S2-4 ventral roots. In surgery, it is very important to determine the innervative frequency and efficacy of every individual nerve root when operation is performed on the sacral roots. We performed those operations in ten cases of Chinese with complete supraconal spinal cord injuries. There were seven men and three women, with an average of 32 years old. Our results show that pressure contributions are provided by S3 roots in all ten patients, bilaterally; and by S4 roots on the right side of ten patients and left side of eight patients. However, only three have minimal pressure uprising by S2 stimulation, one on the right side and two on both sides. The mean innervative efficacy by differential sacral roots to bladder detrusor is shown in Fig. 11.1. S3 root is the most frequent and the most efficacious contributor, S4 is the second frequent and a lesser but still

Table 11.2 Innervations of sacral root nerve to bladder and lower leg muscles

		S2	S3	S4
1.	Detrusor	±	+++	++
2.	Sphincters	+++	++	+±
3.	Sex	+++	±	–
4.	Lower leg muscle	Gluteal major, gultealmedius, femoris biceps, triceps, flexors of hullux and other toes	Triceps, flexors of hullux and other toes	None

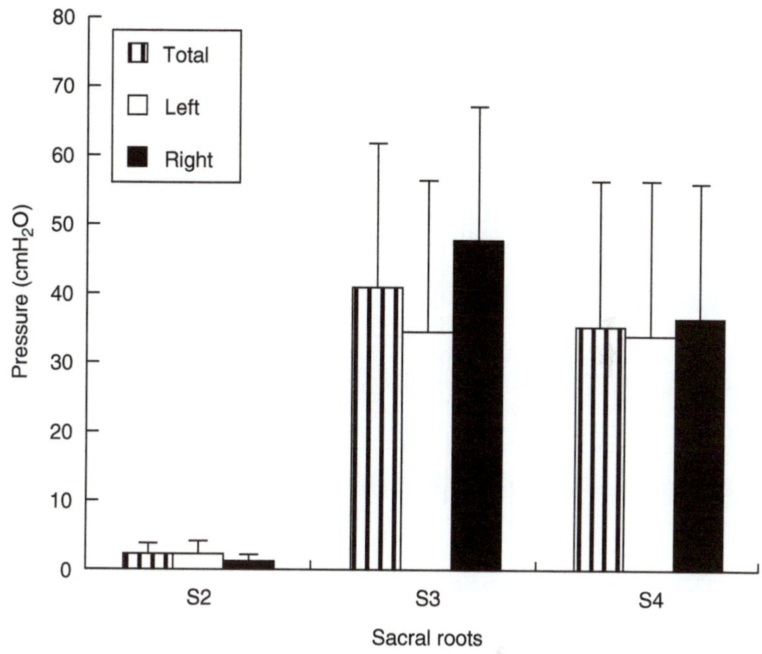

Fig. 11.2 Differential sacral roots innervation to bladder detrusor in Asian people

significant efficacious one, and S2 is the last frequent and the minimal contributor(Fig. 11.2) [20]. The right roots seem to be stronger than the left ones. The efficacy distribution of sacral root innervation to bladder detrusor obtained by us is similar to that by Brindley, but the innervative frequency is different. Rising in bladder pressure is obtainable always by stimulating S3, usually by S4, and often by S2 in Europeans; however, it is always by S3, usually by S4, and seldom by S2 in Asians [20].

The tripolar electrode (arranged with middle cathode and upper and lower anodes) is attached to a strip of silicone rubber (6 mm wide and 39 mm long). One end of this silicone rubber is passed under the appropriate segmental nerve (S2) or nerves (S3 and S4 together). Then the two ends of the rubber strips are sutured. Be careful to check the tension when suturing. The strip of the silicone rubber should be just loose enough to allow a little sliding, avoid nerve entrapment. The upper and lower anodes should be laid within a few millimeters to the nerve. It is not important whether they touch it.

When all the electrodes are in place, the cables are tunneled through to the receiver site, and the block of receivers are implanted.

Another skin incision is made through T12 to L2, and laminectomy was performed to expose conusmadullaris. At this site, it is easy to distinguish anterior rootlets from the posterior rootlets. It is not easy to identify the segmental levels. But anatomical studies shown that, section of all posterior rootlets that enter the last 25 mm of the spinal cord will ensure that the S2 to S5 rootlets are all cut, achieving a complete sacral deafferentation (SDAF) (Figs.11.3 and 11.4).

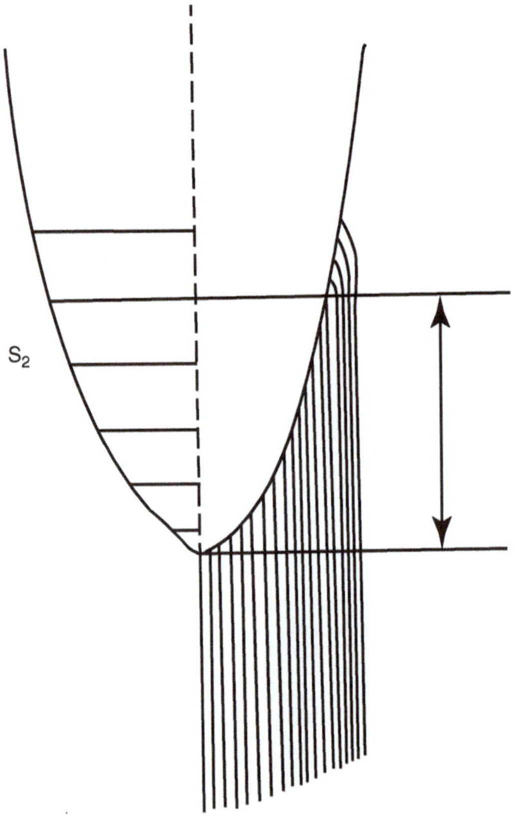

Fig. 11.3 Measuring the distance from terminal dorsal conusmedullaris to the S2 rootlet segment level

11.6 Benefit of Brindley Procedure

1. About 90 % of patients with this implant use it for controlled voiding, and the majority have significant decrease in infection rate and improvement in continence. UTIs dropped from 6.3 per year preoperatively to 1.2 per year postoperatively [21–23].
2. Improvement of bladder capacity to normal and normalisation of compliance; provided that the bladder is no serious fibrosis.
3. Safe continence in most patients following the complete abolition of neurogenic hyperactivity. provided that bladder neck function is intact.
4. Reduction of active detrusor-sphincter -dyssynergia.
5. Over 90 % of the patients considered that they have an improved QOL after implantation of the Brindley stimulator, and would recommend the system to other patients with neurogenic bladder dysfunction.
6. Prevented deterioration of the upper urinary tract by diminution of dilatation, cessation of high pressure ureterorenal reflux and protection of kidney function.

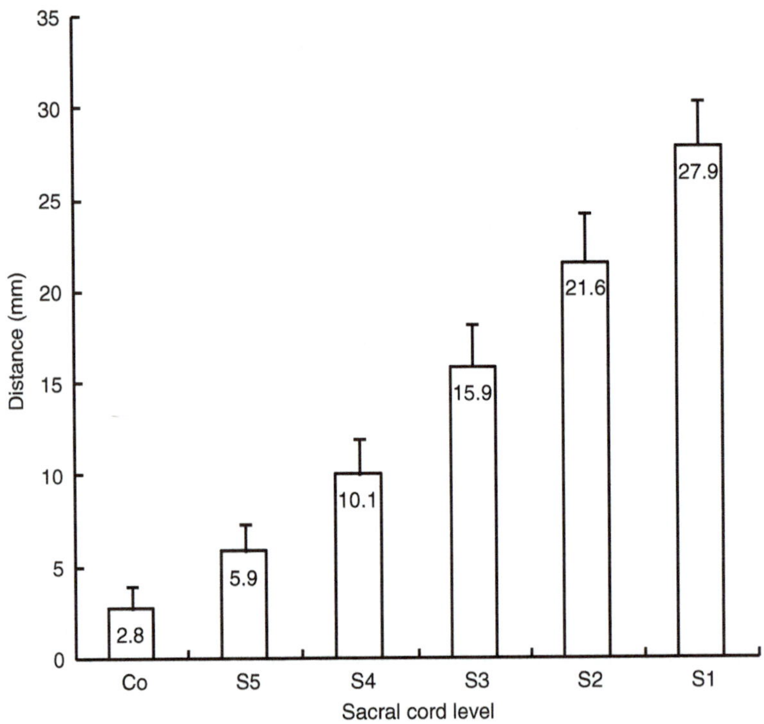

Fig. 11.4 The distance from the terminal conusmedullaris to different sacral segment level

7. In some men, the Brindley System can also promote penile erection.
8. Sacral rhizotomies and electrical bladder stimulation make a cost-effective method of treatment of lower urinary tract dysfunction in patients with SCI. Considerable savings on health care costs are possible in the long-run with simultaneous positive effects on aspects of health status.
9. The Brindley System can eliminate using of urethral catheters, as it can restore satisfactory continence and improve psychological as well as economical constraints related to auto/hetero catheterizations performed several times a day.

11.7 Clinical Outcomes

The Brindley procedure (SARS + SDAF) has been used in more than 3,000 patients. About 20 clinical papers were published in literature [2, 21].

Voiding. The Brindley stimulator is used for micturition in 73–100 % of patients during followup. These are considerable percentages, but it should be noted that this includes patients who may use additional methods to empty their bladder. Additional methods include intermittent catheterization, abdominal straining (Valsalvamanoeuvre), abdominal compression (Cred´emaneuvre), or suprapubic tapping for reflex contractions. Stimulation that is not always completely successful can be found back in the percentages of patients that have less than 50 mL residual urine after stimulation for micturition. These percentages are lower than the percentages of patients that use the stimulator for micturition. Overall, the percentages of patients having urinary tract infections and the frequency of urinary tract infections decrease after the Brindley procedure compared to the preoperative treatment.

Defecation. TheBrindley stimulator is used for defecation in 29–100 % of patients in different degrees. Not all patients achieve complete evacuation of defecation using only stimulation. Some patients need laxatives in addition to prevent constipation or enable defecation. Many patients only use the stimulator to get the defecation into the rectum, to enable digital evacuation.

Erection. Erections can be evoked in a substantial number of patients, but results vary considerably. This can be explained by the relatively low number of patients that actually use the stimulator to evoke erections for sexual intercourse (0–32 %), due to qualitatively inadequate erections for sexual intercourse or deterioration of the stimulation effect over time.

Autonomicdysreflexia. Autonomicdysreflexia mostly decreased after the Brindley procedure as a result of the dorsal rhizotomy. Only a few studies reported stimulation-induced autonomic dysreflexia.

Continence. Continence is achieved in 57–100 % of patients, and bladder capacity increased. However, continence is not only achieved by a dorsal rhizotomy. Results on continence also included additional treatments, like anticholinergics and stress incontinence surgery.

11.8 Illustrative Case

A 25-year-old female suffered with T8 fracture and paraplegia for 3 years. She was performed SARS-SDAF in 1998 with the Barcelona procedure (Fig. 11.5).

11.9 Problems and Future Investigations

The ultimate treatment of neurogenic disorders of the lower urinary tract would be resolvement of the neurogenic disorder that causes the bladder problems to restore the innervation of the bladder. As long as this causal treatment is not available, symptomatic treatment options are required.

The Brindley procedure generally shows good clinical results for restoration of function in spinal cord injury patients with multiple pelvic organ dysfunction, including bladder, bowel, and erectile dysfunction. Moreover, the Brindley procedure improves quality of life.

However, it is not a procedure that is easy to apply in clinical practice:

1. Not every patient is suited for the procedure and the success depends on selection of appropriate patients. Prerequisites are a complete spinal cord lesion since neurostimulation can cause pain in incomplete spinal cord lesions,

an intact sacral motor neuron pathway enabling stimulation of the bladder, and a detrusor muscle that is capable to contract on stimulation.

2. A dorsal rhizotomy and implantation of a Brindley stimulator is complex and not a routine procedure for urologists and should be reserved for specialized centers.
3. The device is also prone to failures, including the external and implanted components. Analysis of the external components is easy to apply. Currently, a straightforward solution for analysis and revision of the implanted system without major surgery is not available.
4. The neuroprosthesis is expensive for patients living in developing countries. We propose a modification to make it cheaper and simpler: using one cable with two electrodes to trap on the bilateral S3 and S4 root nerves combined with cutting distal 20 mm conusmedullaris as de-afferentation (Fig. 11.6). This can make patients get both voiding and storing function, but no penile erection [24, 25].

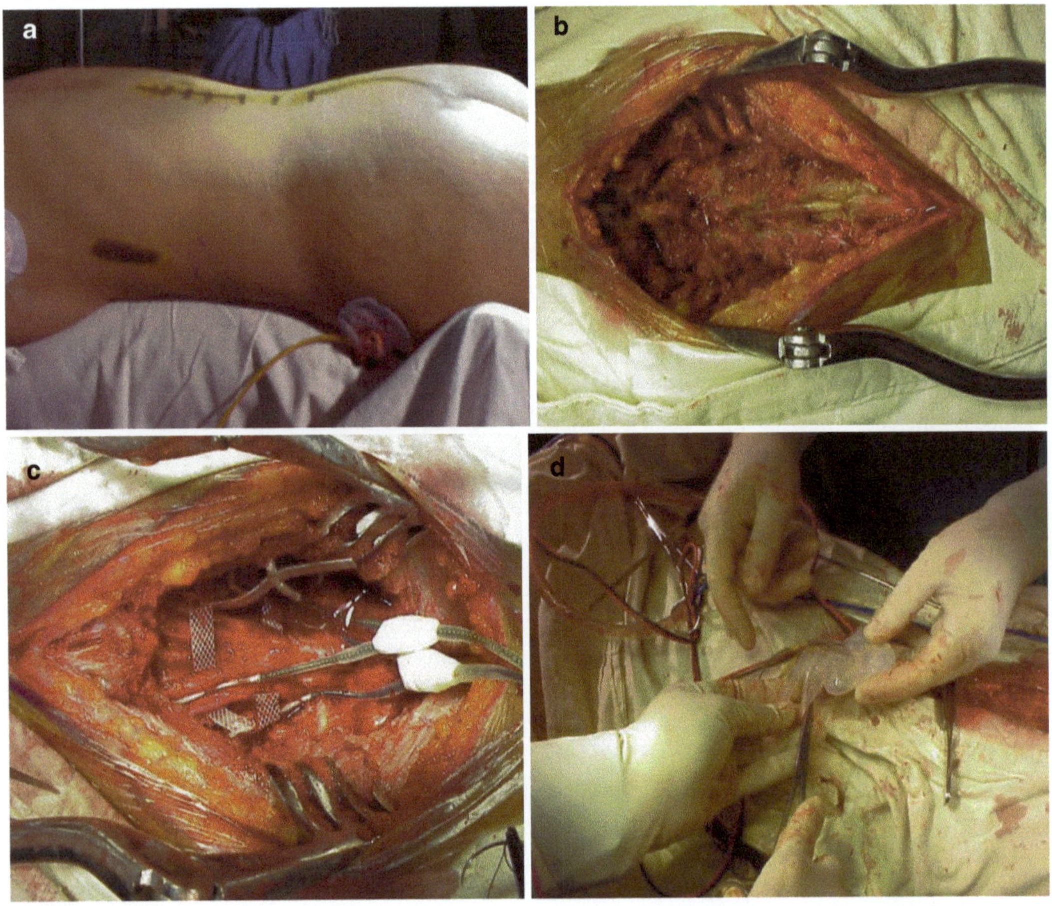

Fig. 11.5 Surgical technique (**a**) The patient was placed in prone position and three skin incisions were marked. (**b**) Sacral skin exposure of the laminar. (**c**) After laminectomy, sacral root nerves were exposed. Two bundles (each with two cables attached with two electrodes) were used to trap bilateral S2 and S3/4 separately. (**d**) The cables were passed to the lateral thoracic incision, and connected to the receive block. (**e**) The receive block was inserted under the subcutaneous tissue and sutured onto the deep fascia. (**f**) Sacral de-afferentation was performed at the conusmedullaris level, by cutting the terminal 25 mm dorsal rootlets bilaterally. (**g**) Three skin incisions were closed. (**h**) External stimulator was put onto the implanted receive block of the lateral thoracic site. (**i**) Electrical controlled voiding was very robust in followup

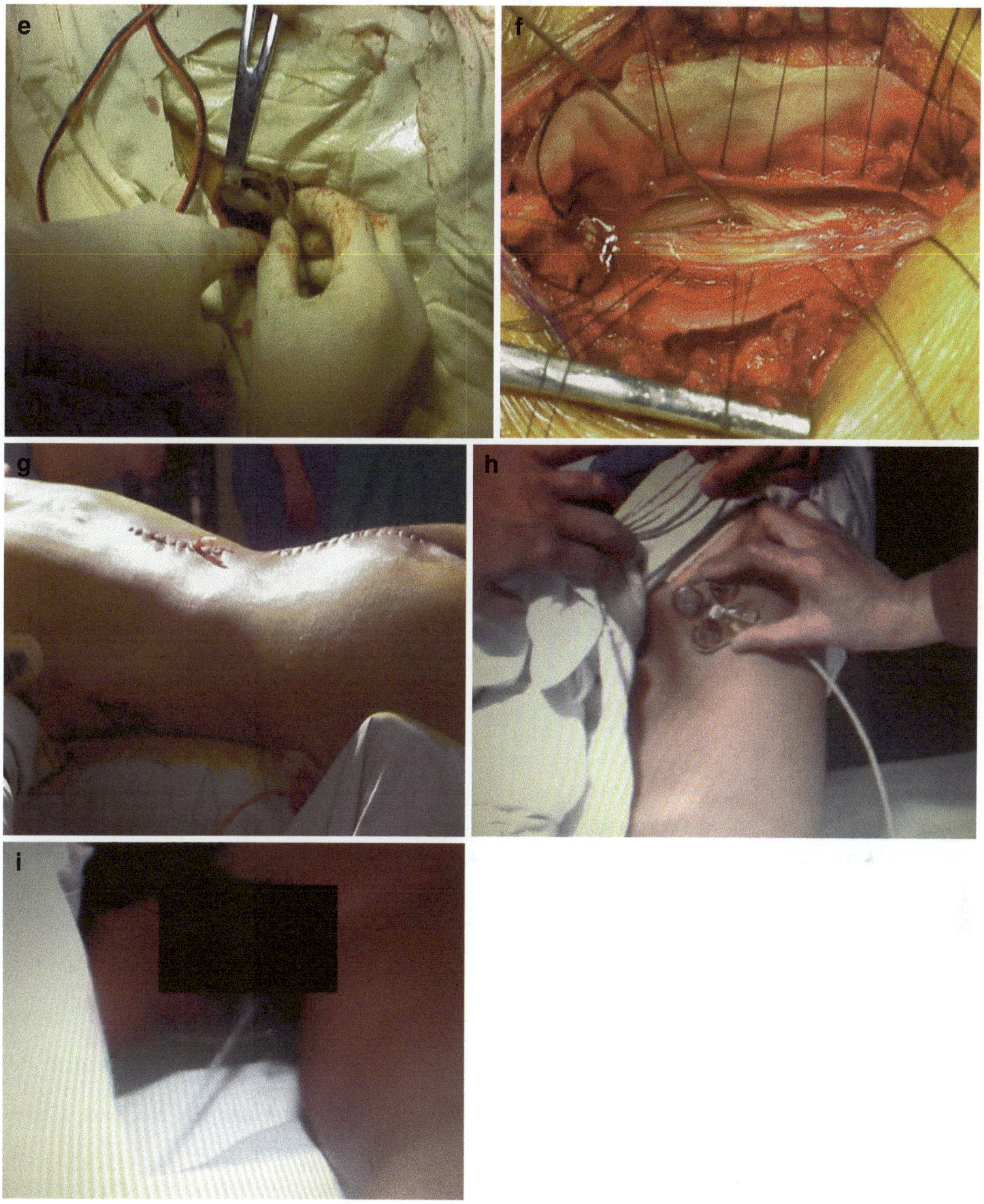

Fig. 11.5 (continued)

5. A main issue for patients who consider a Brindley procedure is the irreversibility of the rhizotomy, and the possibility that future treatment options are not within reach anymore. Staggered rhizotomy of anterior and posterior sacral root to leave the S2 posterior root intact has been investigated in animal models [26]. Continuous or conditional neuromodulation with low voltage using the stimulator could be one of the solutions [23]. To avoid detrusor-external sphincter dyssynergia, selective detrusor stimulation, such like selective anodal block and high-frequency block for the somatic nerve fibers, have been investigated [27–31].

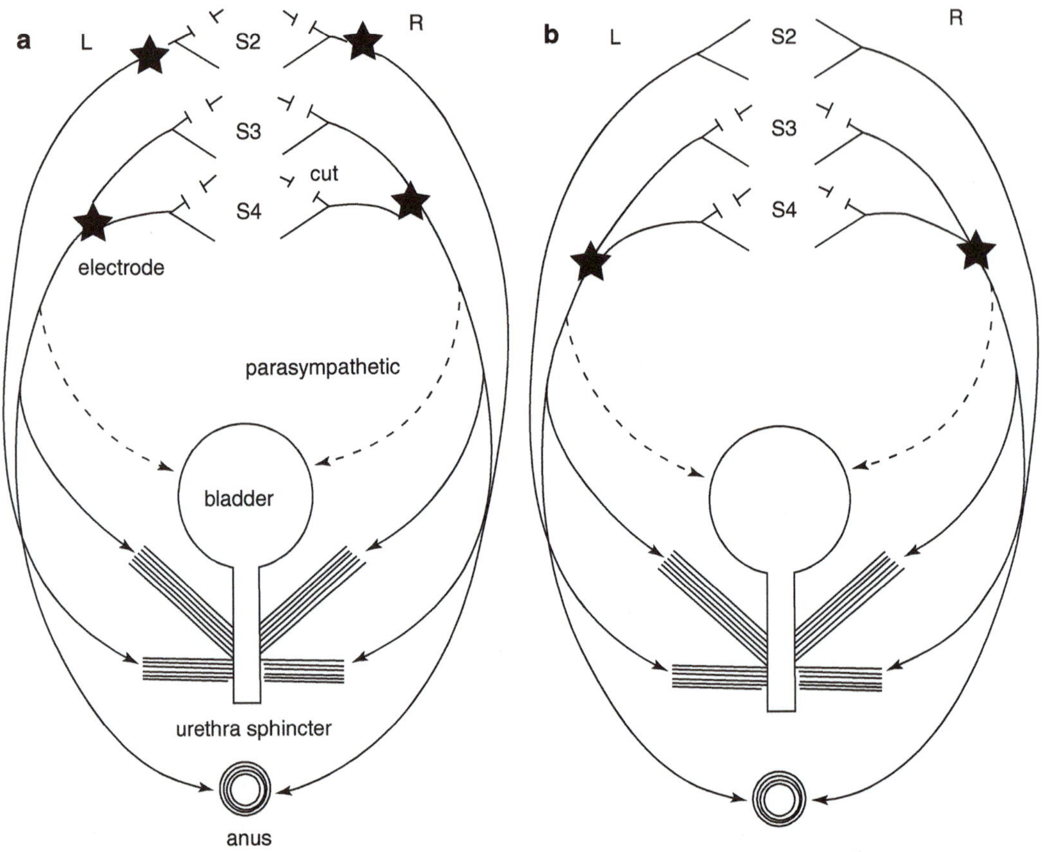

Fig. 11.6 (**a**) Standard method: 4 cables-4 electrodes to stimulate bilateral S2 and bilateral S3/4. (**b**) Modified simpler method: only two cables with two electrodes are used to stimulate bilateral S3/4

Conclusion

The Brindley procedure shows good clinical results and improves quality of life in spinal cord injured patients. However, to remain a valuable treatment option for the future, the technique still needs some changes, for example to abolish the sacral dorsal rhizotomy, and to lower the price.

References

1. Brindley GS, Polkey CE, Rushton DN, Cardozo L. Sacral anterior root stimulators for bladder control in paraplegia: the first 50 cases. J Neurol Neurosurg Psychiatry. 1986;49:1104–14.
2. Brindley GS. The first 500 patients with sacral anterior root stimulator implants: general description. Paraplegia. 1994;32:795–805.
3. Sauerwein D. Surgical treatment of spastic bladder paralysis in paraplegic patients. Sacral deafferentation with implantation of a sacral anterior root stimulator. Urologe A. 1990;29:196–203.
4. Hohenfellner M, Pannek J, Bötel U, et al. Sacral bladder denervation for treatment of detrusor hyperreflexia and autonomic dysreflexia. Urology. 2001;58:28–32.
5. Van Kerrebroeck PE, Koldewijn EL, Debruyne FM. Worldwide experience with the Finetech-Brindley sacral anterior root stimulator. Neurourol Urodyn. 1993;12:497–503.
6. Van Kerrebroeck PEV, Koldewijn EL, Rosier PF, Wijkstra H, Debruyne FMJ. Results of the treatment of neurogenic bladder dysfunction in spinal cord injury by sacral posterior root rhizotomy and anterior sacral root stimulation. J Urol. 1996;155:1378–81.
7. Creasey GH, Grill JH, Korsten M, et al. An implantable neuroprosthesis for restoring bladder and bowel control to patients with spinal cord injuries: a multicenter trial. Arch Phys Med Rehabil. 2001;82:1512–9.
8. Egon G, Barat M, Colombel P, Visentin C, Isambert JL, Guerin J. Implantation of anterior sacral root stimulators combined with posterior sacral rhizotomy in spinal injury patients. World J Urol. 1998;16:342–9.

9. Madersbacher H, Fischer J. Sacral anterior root stimulation: prerequisites and indications. Neurourol Urodyn. 1993;12:489–94.
10. Hamel O, Perrouin-Verbe B, Robert R. Brindley technique with intraduraldeafferentation and extradural implantation by a single sacral laminectomy. La Neurochirurgie. 2004;50:661–6.
11. Kutzenberger J, Domurath B, Sauerwein D. Spastic bladder and spinal cord injury: seventeen years of experience with sacral deafferentation and implantation of an anterior root stimulator. Artif Organs. 2005;29:239–41.
12. Macdonagh RP, Sun WM, Smallwood R, Forster D, Read NW. Control of defecation in patients with spinal injuries by stimulation of sacral anterior nerve roots. Br Med J. 1990;300:1494–7.
13. Madersbacher H, Fischer J, Ebner A. Anterior sacral root stimulator (Brindley): experiences especially in women with neurogenic urinary incontinence. Neurourol Urodyn. 1988;7:593–601.
14. Robinson LQ, Grant A, Weston P, Stephenson TP, Lucas M, Thomas DG. Experience with the Brindley anterior sacral root stimulator. Br J Urol. 1988;62: 553–7.
15. Sauerwein D, Ingunza W, Fischer J, et al. Extradural implantation of sacral anterior root stimulators. J Neurol Neurosurg Psychiatry. 1990;53:681–4.
16. Schurch B, Rodic B, Jeanmonod D. Posterior sacral rhizotomy and intradural anterior sacral root stimulation for treatment of the spastic bladder in spinal cord injured patients. J Urol. 1997;157:610–4.
17. van der Aa HE, Alleman E, Nene A, Snoek G. Sacral anterior root stimulation for bladder control: clinical results. Arch Physiol Biochem. 1999;107:248–56.
18. Sarrias M, Sarrias F, Borau A. The "Barcelona" technique. Neurourol Urodyn. 1993;12:495–6.
19. Vastenholt JM, Snoek GJ, Buschman HPJ, Van Der Aa HE, Alleman ERJ, Ijzerman MJ. A 7-year follow-up of sacral anterior root stimulation for bladder control in patients with a spinal cord injury: quality of life and users' experiences. Spinal Cord. 2003;41: 397–402.
20. Chang SM, Hou CL. The frequency and efficacy of differential sacral roots innervation to bladder detrusor in Asian people. Spinal Cord. 2000;38:773.
21. Martens FM, den Hollander PP, Snoek GJ, Koldewijn EL, van Kerrebroeck PE, Heesakkers JP. Quality of life in complete spinal cord injury patients with a Brindley bladder stimulator compared to a matched control group. Neurourol Urodyn. 2011;30:551–5.
22. Hansen J, Media S, Nøhr M, Biering-Sørensen F, Sinkjaer T, Rijkhoff NJM. Treatment of neurogenic detrusor overactivity in spinal cord injured patients by conditional electrical stimulation. J Urol. 2005;173: 2035–9.
23. Kirkham APS, Knight SL, Craggs MD, Casey ATM, Shah PJR. Neuromodulation through sacral nerve roots 2 to 4 with a Finetech-Brindley sacral posterior and anterior root stimulator. Spinal Cord. 2002;40: 272–81.
24. Chang SM, Hou CL, Wang SB, et al. Sacral anterior root stimulated defecation in spinal cord injuries: an experimental study in canine model. World J Gastroenterol. 2005;11:1715–8.
25. Chang SM, Hou CL, Xu RS. Applied anatomy of complete deafferentation at conusmedullaris for electrical stimulated micturition in Asian spinal cord injury patients. J Clin Anat. 2002;1:50–2.
26. Chang SM, Hou CL. Staggered rhizotomy of anterior and posterior sacral root for Bladder reservoir function in spinal cord injuries: canine experimental study and preliminary clinical report. Orthop Surg. 2009;1:164–70.
27. Boger A, Bhadra N, Gustafson KJ. Bladder voiding by combined high frequency electrical pudendal nerve block and sacral root stimulation. Neurourol Urodyn. 2008;27:435–9.
28. Brindley GS, Craggs MD. A technique for anodally blocking large nerve fibres through chronically implanted electrodes. J Neurol Neurosurg Psychiatry. 1980;43:1083–90.
29. Rijkhoff NJ, Wijkstra H, Van Kerrebroeck PE, Debruyne FM. Selective detrusor activation by electrical sacral nerve root stimulation in spinal cord injury. J Urol. 1997;157:1504–8.
30. Seif C, Braun PM, Bross S, et al. Selective block of urethral sphincter contraction using a modified brindley electrode in sacral anterior root stimulation of the dog. Neurourol Urodyn. 2002;21:502–10.
31. Martens FMJ, Heesakkers JPFA. Brindley sacral anterior-root stimulator (an implantable neuroprosthesis) for restoring bladder and bowel control to patients with spinal cord injuries. Adv Urol. 2011. Article ID 709708

Index

C. Hou (ed.), *Functional Bladder Reconstruction Following Spinal Cord Injury via Neural Approaches*, DOI 10.1007/978-94-007-7766-8, © Springer Science+Business Media Dordrecht 2014

FSC
www.fsc.org
MIX
Papier | Fördert
gute Waldnutzung
FSC® C083411

Kolibri 360 GmbH
Ferdinand-Jühlke-Straße 7
99095 Erfurt, Deutschland
produktsicherheit@kolibri360.de